Numerical Techniques for Direct and Large-Eddy Simulations

T0225587

CHAPMAN & HALL/CRC
Numerical Analysis and Scientific Computing

Aims and scope:
Scientific computing and numerical analysis provide invaluable tools for the sciences and engineering. This series aims to capture new developments and summarize state-of-the-art methods over the whole spectrum of these fields. It will include a broad range of textbooks, monographs, and handbooks. Volumes in theory, including discretisation techniques, numerical algorithms, multiscale techniques, parallel and distributed algorithms, as well as applications of these methods in multi-disciplinary fields, are welcome. The inclusion of concrete real-world examples is highly encouraged. This series is meant to appeal to students and researchers in mathematics, engineering, and computational science.

Editors

Proposals for the series should be submitted to one of the series editors above or directly to:
CRC Press, Taylor & Francis Group
4th, Floor, Albert House
1-4 Singer Street
London EC2A 4BQ
UK

Published Titles

A Concise Introduction to Image Processing using C++
Meiqing Wang and Choi-Hong Lai

**Decomposition Methods for Differential Equations:
   Theory and Applications**
Juergen Geiser

**Grid Resource Management: Toward Virtual and Services Compliant
Grid Computing**
Frédéric Magoulès, Thi-Mai-Huong Nguyen, and Lei Yu

Introduction to Grid Computing
Frédéric Magoulès, Jie Pan, Kiat-An Tan, and Abhinit Kumar

Numerical Linear Approximation in C
Nabih N. Abdelmalek and William A. Malek

Numerical Techniques for Direct and Large-Eddy Simulations
Xi Jiang and Choi-Hong Lai

Parallel Algorithms
Henri Casanova, Arnaud Legrand, and Yves Robert

Parallel Iterative Algorithms: From Sequential to Grid Computing
Jacques M. Bahi, Sylvain Contassot-Vivier, and Raphael Couturier

Numerical Techniques for Direct and Large-Eddy Simulations

Xi Jiang
Choi-Hong Lai

CRC Press
Taylor & Francis Group
Boca Raton   London   New York

CRC Press is an imprint of the
Taylor & Francis Group, an **informa** business

A CHAPMAN & HALL BOOK

CRC Press
Taylor & Francis Group
6000 Broken Sound Parkway NW, Suite 300
Boca Raton, FL 33487-2742

First issued in paperback 2017

© 2009 by Taylor and Francis Group, LLC
CRC Press is an imprint of Taylor & Francis Group, an Informa business

No claim to original U.S. Government works

ISBN 13: 978-1-138-11383-1 (pbk)
ISBN 13: 978-1-4200-7578-6 (hbk)

Library of Congress Cataloging-in-Publication Data

Jiang, Xi, 1967-
    Numerical techniques for direct and large-eddy simulations / Xi Jiang, Choi-Hong Lai.
        p. cm. -- (Chapman & Hall/CRC numerical analysis and scientific computing)
    Includes bibliographical references and index.
    ISBN 978-1-4200-7578-6 (hardcover : alk. paper)
    1. Eddies--Mathematical models. 2. Turbulence--Mathematical models. 3. Fluid dynamics--Mathematical models. I. Lai, Choi-Hong. II. Title. III. Series.

TA357.5.T87J53 2009
620.1'064015118--dc22
                                                        2009018492

Visit the Taylor & Francis Web site at
http://www.taylorandfrancis.com

and the CRC Press Web site at
http://www.crcpress.com

To Our Families

Table of Contents

Preface

Numerical Techniques for Direct and Large-Eddy Simulations intends to build up a text suitable to be used as lecture material for postgraduate students and a reference for research scientists and engineers interested in advanced numerical simulations of fluid flow problems and the associated numerical methods. Direct numerical simulation (DNS) and large-eddy simulation (LES) are advanced numerical tools for computational fluid dynamics (CFD), which have developed enormously over the last few decades. In comparison to traditional CFD based on the Reynolds-averaged Navier–Stokes (RANS) modeling approach, DNS and LES offer much enhanced capability in predicting the unsteady features of the flow field, such as the vortical structures, and provide very detailed solution of the flow field that can be used not only to obtain a better understanding but also to develop models for mixing and turbulence. DNS and LES aim to simulate fluid flows with high fidelity using high-resolution numerical methods. In many cases, DNS can obtain results that are not possible using any other means, while LES is being adopted by industry as an advanced tool for practical applications.

The main themes of this book are the numerical needs arising from applications of DNS and LES. This book covers the basic techniques for DNS and LES that can be applied to practical problems in the field of flow, turbulence, and combustion. The book includes specific numerical techniques for compressible and incompressible flows. It focuses on numerical methods that are suitable to three-dimensional flows, rather than specific numerical methods for one- or two-dimensional flows. The book is intended to provide fundamental knowledge on numerical techniques for DNS and LES—for example, high-order discretization schemes, high-fidelity boundary conditions, and coupling aspects—in order to help the readers understand the details of the numerical methods used in DNS and LES and the substantial amount of literature in the field. The book is limited to explaining the numerical techniques concisely so that most of the relevant numerical techniques for DNS and LES can be discussed. Of particular interest in the text are the sample numerical simulation results given in the relevant chapters, which exemplify the practical applications of the numerical techniques under discussion.

The text has been organized in a way that is easy to understand for postgraduate students with a basic knowledge of CFD. In fact, most of the materials included in this book have evolved from our lecture notes and other teaching materials. The numerical techniques presented are not intended as thorough descriptions of the methods concerned; rather they are intended to help CFD practitioners in the early stage of their academic careers, such as postgraduate students and other junior researchers, to understand the vast amount of literature in the field published mainly in academic journals and, more importantly, to apply the relevant numerical techniques in practical CFD simulations or to implement these methods in their CFD computer programs. More detailed and more in-depth information on the numerical techniques are found in the references cited.

In this book, the individual chapters are intended as concise descriptions of the relevant areas related to the numerical aspects of direct and large-eddy simulations, including numerical methods for both incompressible and compressible flows. The numerical methods for compressible flows discussed in the text are mainly restricted to relatively low-speed compressible flows without significant formation of shock waves. Chapter 1 presents an introduction to the Navier–Stokes equations and the methodologies of DNS and LES. Chapter 2 discusses boundary conditions for DNS and LES. Chapter 3 presents the time integration methods. Chapter 4 describes the numerical techniques used in DNS of incompressible flows. Chapter 5 describes the numerical techniques used in DNS of compressible flows. Chapter 6 describes the basic LES techniques for simulating incompressible flows. Chapter 7 covers LES of compressible flows. Chapter 8 is devoted to further topics and current challenges in DNS and LES. In Chapters 4–8, sample numerical simulation results are included. Finally, sample FORTRAN programs are included in the appendix to illustrate the implementation of finite difference numerical schemes.

Acknowledgments

Xi Jiang and Choi-Hong Lai are grateful to many of their colleagues and students, with whom they have worked together in the field of advanced computational fluid dynamics. *Numerical Techniques for Direct and Large-Eddy Simulations* was mainly built upon the lecturing materials for advanced thermofluids and computational fluid dynamics in the authors' host institutions. Numerous discussions were held with the departmental colleagues and students taking the modules. Their useful comments and feedback were extremely helpful in the compilation of the text.

We would like to express our deepest gratitude to George Siamas, who worked closely with Xi Jiang at Brunel University during the writing of this book. Most of the examples included in the book were selected by Dr. Siamas. In addition, George Siamas checked some contents of the manuscript and put together the FORTRAN program in the appendix. Xi Jiang is also extremely grateful to Kai H. Luo at the University of Southampton, whose constant support, inspiration, and useful advice have helped the most throughout his research in the field of direct and large-eddy simulations. The research of Xi Jiang in the field has been supported by the UK Turbulence Consortium and the Consortium on Computational Combustion for Engineering Applications with valuable high-performance computing resources, which are gratefully acknowledged.

Last, but most important, we would like to thank our families, who have provided us with constant encouragement, unconditional support, and enormous inspiration. Without their understanding and sacrifice, it would not be possible for us to accomplish this text and any other academic progress.

Xi Jiang

Choi-Hong Lai

Authors

Xi Jiang is currently senior lecturer in the subject area of Mechanical Engineering, School of Engineering and Design, at Brunel University in the United Kingdom. In 1994, Jiang obtained his PhD degree from the University of Science and Technology of China. In that year, he was the winner of the Guo Mo-ruo Award of the university. Since then, he has been doing research at the University of Science and Technology of China; Seoul National University; Fire Research Station, Building Research Establishment (United Kingdom); Queen Mary, University of London; and Brunel University. He has more than fifteen years successful experience in numerical studies of fluid flow and combustion. In 2002, Jiang (together with Professor Luo at the University of Southampton) was awarded the Gaydon Prize for the most significant UK contribution to the 28th Symposium (International) on Combustion (Edinburgh, 2000). Over the last several years, he has been working on various areas of numerical simulation and modeling of turbulence, combustion, and aeroacoustics. The physical problems investigated by him covered a broad range of practical applications, mainly reacting and nonreacting jets and plumes, gas-liquid two-phase jet flows, gas-solid two-phase reactive flows, flame spread in microgravity environments, and noise generation from subsonic jets. In total, he has published more than 40 journal papers.

Jiang started his research on direct numerical simulation (DNS) of flow and combustion in 1998, as a research fellow with K.H. Luo at Queen Mary, University of London (now at the University of Southampton). Since then he has been continuously using the UK national high-performance computing (HPC) resources for his research work on numerical studies of fluid flow and combustion. Based on the HPC resources, his on-going research is focused on DNS of nonpremixed flames and nonreacting gas-liquid two-phase jet flows and large-eddy simulation (LES) of fuel injection and spray combustion. Jiang is currently a member of the UK Turbulence Consortium, which is working on the package "multiphase phenomena." He is also a member of the Consortium on Computational Combustion for Engineering Applications.

Choi-Hong Lai is professor of numerical mathematics in the Scientific Computing and Algorithms Group within the Centre for Numerical Modelling and Process Analysis (CNMPA), School of Computing and Mathematical Sciences, at the University of Greenwich in the United Kingdom. Lai joined the University in 1989 after a PhD in parallel and distributed algorithms for fluid dynamics and partial differential equations and four years of postdoctoral experience in parallel finite element methods at Queen Mary, University of London. He has been involved in parallel and distributed algorithm development employing the concept of domain decomposition with applications in aeroacoustics, coupling of numerical models, and defect correction methods. His research interests include numerical methods for partial differential equations, scientific computing, coupling of multiphysics mathematical models and heterogeneous numerical methods, and high-order schemes for computational aeroacoustics. Recently, Lai and his research team at CNMPA started working in the area of high-order schemes for computational fluid dynamics applications, particularly for direct and large-eddy simulations. Lai is very experienced in the development of robust numerical algorithms for science and engineering applications. He has published over 40 journal papers and is the computer mathematics subject editor for the *International Journal of Computer Mathematics* and the editor for the *Journal of Algorithms and Computational Technology*.

CHAPTER 1

Introduction

A S AN INTERDISCIPLINARY SCIENCE, CFD is now applied in the fields of aerospace, automotive, biomedical, civil, chemical, environmental, mechanical, and even electrical engineering, as well as in physics, chemistry, and biology. As Anderson (1995) stated, modern CFD cuts across all disciplines where the flow of a fluid is playing a significant role. CFD is essentially a science of replacing the governing equations describing the fundamental physical principles with discretized algebraic forms, which in turn are solved to obtain numerical values for the flow field at discrete points in space and/or in time. This book is focused on the numerical methods to discretize the governing equations. However, it is necessary to introduce the governing equation before discussing the numerical methods.

All methods of CFD, whether the traditional RANS approach or other advanced approaches such as DNS or LES, are built upon the fundamental governing equations of fluid dynamics—the continuity equation based on mass conservation, the momentum equations based on Newton's second law of motion, and the energy equation based on energy conservation. For combustion applications and many other applications involving chemical reactions, there are also governing equations for mass conservation of different species of chemicals. This chapter is intended as an introduction to the governing equations (Navier–Stokes equations) and the methodologies of DNS and LES. Due to the significant differences between the numerical methods for compressible and incompressible flows, discussions of numerical features for compressible and incompressible flows are also included when the governing equations are presented.

I. GOVERNING EQUATIONS: COMPRESSIBLE AND INCOMPRESSIBLE FORMULATIONS

A. Fundamental Governing Equations for Fluid Flows

The governing equations for fluid flow are the mathematical statements of three fundamental physical principles upon which all motions of fluid are based:

- Mass is conserved—the continuity equation.

- Newton's second law $(\mathbf{F} = m\mathbf{a})$ —the momentum equation.

- Energy is conserved—the energy equation.

The governing equations concern the physics of fluid. Both theoretical and computational fluid dynamics are based on these equations, and therefore it is essential for a CFD practitioner to be familiar with them and to understand their physical significance. Using an intuitive and physically oriented approach, Anderson (1995) provided an excellent description of the governing equations for CFD suitable for readers who are inexperienced in CFD.

In order to obtain the basic equations of fluid motion at the macroscale, such as meters or millimeters, the following procedure is always followed:

- First, choose the appropriate fundamental physical principles from the law of physics, such as (1) mass conservation, (2) Newton's second law, and (3) energy conservation.

- Second, apply these physical principles to a suitable model of the flow. Unlike a solid body, which is easy to see and to define and has a definite shape, a fluid is a substance that is hard to grab hold of and does not have a shape unless a container is used. If a fluid is in motion, the velocity may be different at each location in the fluid. Therefore, flow models are needed to identify the moving fluid. For a continuum fluid, there are four possible models that can be used to visualize the moving fluid so as to apply to it the fundamental physical principles: (1) finite control volume fixed in space with the fluid moving through it, (2) finite control volume moving with the fluid such that the same fluid particles are always in the same control volume, (3) infinitesimal fluid element fixed in space with the fluid moving through it, and (4) infinitesimal fluid element moving along a streamline with the velocity equal to the local flow velocity at each point.

- Third, apply a physical principle to a particular flow model and extract the mathematical equations that embody such physical principles.

Applying the fundamental physical principles to one of the four models will result in the basic governing equations for fluid dynamics: continuity, momentum, and energy equations. In the finite control volume approach, the fundamental physical principles are applied to the fluid inside the control volume. The flow equations directly obtained are in integral form. The integral form of the governing equations can be manipulated to indirectly obtain a set of partial differential equations. The approach of an infinitesimal fluid element, in the meantime, leads directly to the fundamental equations in a partial differential equation form. Similarly, the partial differential equation form of the governing equations can be manipulated to obtain the respective integral equations. In the following, the partial differential form of the governing equations based on the approach of infinitesimal fluid element is examined. The fluid element is "infinitesimal" in the same sense as in differential calculus; however, it is large enough to contain a huge number of molecules so that it can be viewed as a continuous medium.

Using the approach of an infinitesimal fluid element moving along a streamline and considering a fluid element moving through Cartesian space with the velocity $\mathbf{V} = u\mathbf{i} + v\mathbf{j} + w\mathbf{k}$, where $\mathbf{i}$, $\mathbf{j}$, and $\mathbf{k}$ are the unit vectors along the x, y, and z axes respectively, the fundamental governing equations for an unsteady, three-dimensional, compressible, and viscous flow can be written as follows:

The continuity equation,

$$\frac{\partial \rho}{\partial t} + \nabla \cdot (\rho \mathbf{V}) = 0 \tag{1.1}$$

Momentum equations (the Navier–Stokes equations)
x component,

$$\frac{\partial(\rho u)}{\partial t} + \nabla \cdot (\rho u \mathbf{V}) = -\frac{\partial p}{\partial x} + \frac{\partial \tau_{xx}}{\partial x} + \frac{\partial \tau_{yx}}{\partial y} + \frac{\partial \tau_{zx}}{\partial z} + \rho f_x \tag{1.2}$$

y component,

$$\frac{\partial(\rho v)}{\partial t} + \nabla \cdot (\rho v \mathbf{V}) = -\frac{\partial p}{\partial y} + \frac{\partial \tau_{xy}}{\partial x} + \frac{\partial \tau_{yy}}{\partial y} + \frac{\partial \tau_{zy}}{\partial z} + \rho f_y \tag{1.3}$$

z component,

$$\frac{\partial(\rho w)}{\partial t} + \nabla \cdot (\rho w \mathbf{V}) = -\frac{\partial p}{\partial z} + \frac{\partial \tau_{xz}}{\partial x} + \frac{\partial \tau_{yz}}{\partial y} + \frac{\partial \tau_{zz}}{\partial z} + \rho f_z \qquad (1.4)$$

The energy equation,

$$\frac{\partial}{\partial t}\left[\rho\left(e + \frac{V^2}{2}\right)\right] + \nabla \cdot \left[\rho\left(e + \frac{V^2}{2}\right)\mathbf{V}\right]$$

$$= \rho \dot{q} + \frac{\partial}{\partial x}\left(k\frac{\partial T}{\partial x}\right) + \frac{\partial}{\partial y}\left(k\frac{\partial T}{\partial y}\right) + \frac{\partial}{\partial z}\left(k\frac{\partial T}{\partial z}\right)$$

$$-\frac{\partial(u p)}{\partial x} - \frac{\partial(v p)}{\partial y} - \frac{\partial(w p)}{\partial z} + \frac{\partial(u \tau_{xx})}{\partial x} + \frac{\partial(u \tau_{yx})}{\partial y} + \frac{\partial(u \tau_{zx})}{\partial z}$$

$$+\frac{\partial(v \tau_{xy})}{\partial x} + \frac{\partial(v \tau_{yy})}{\partial y} + \frac{\partial(v \tau_{zy})}{\partial z} + \frac{\partial(w \tau_{xz})}{\partial x} + \frac{\partial(w \tau_{yz})}{\partial y} + \frac{\partial(w \tau_{zz})}{\partial z} + \rho \mathbf{f} \cdot \mathbf{V}$$

$$(1.5)$$

In the above governing equations, variables have their usual meanings: e stands for the internal energy per unit mass; $\mathbf{f}$ the body forces acting on the volumetric mass of the fluid element (such as the gravitational, electric, and magnetic forces); k the thermal conductivity; p the pressure; $\dot{q}$ the rate of volumetric heat addition (such as combustion heat release, absorption, or emission of radiation) per unit mass; u, v, and w the velocity components in the x, y, and z directions with $V^2 = u^2 + v^2 + w^2$; t time; T the temperature; ρ the density; and τ the viscous stresses, and $\nabla \equiv \mathbf{i}\,\partial/\partial x + \mathbf{j}\,\partial/\partial y + \mathbf{k}\,\partial/\partial z$ is the vector operator in Cartesian coordinates. Among the variables in the governing equations, the shear and normal stresses in a fluid are related to the temporal rate of change of the deformation of the fluid element. The shear stress is related to the temporal rate of change of the shearing deformation of the fluid element. By convention, τ_{ij} denotes a stress in the j direction exerted on a plane perpendicular to the i axis. The normal stress is related to the temporal rate of change of volume of the fluid element, which is usually smaller than the shear stress. In the late seventeenth century, Isaac Newton stated that shear stress in a fluid is proportional to the time rate of strain (i.e., velocity gradients). Such fluids are called Newtonian fluids. In many fluid dynamic problems for engineering applications, the fluid can be assumed to be Newtonian. For such fluids, Stokes in 1845 obtained

$$\tau_{xy} = \tau_{yx} = \mu\left(\frac{\partial v}{\partial x} + \frac{\partial u}{\partial y}\right), \quad \tau_{xz} = \tau_{zx} = \mu\left(\frac{\partial w}{\partial x} + \frac{\partial u}{\partial z}\right),$$

$$\tau_{yz} = \tau_{zy} = \mu\left(\frac{\partial w}{\partial y} + \frac{\partial v}{\partial z}\right), \quad \tau_{xx} = \lambda(\nabla \cdot \mathbf{V}) + 2\mu\frac{\partial u}{\partial x},$$

$$\tau_{yy} = \lambda(\nabla \cdot \mathbf{V}) + 2\mu\frac{\partial v}{\partial y}, \quad \tau_{zz} = \lambda(\nabla \cdot \mathbf{V}) + 2\mu\frac{\partial w}{\partial z} \tag{1.6}$$

where μ is the molecular viscosity coefficient and λ is the second viscosity coefficient. Stokes made the hypothesis that $\lambda = -2/3\,\mu$.

Non-Newtonian fluids are encountered in many other applications such as biological flows. These are fluids that depart from the classic linear Newtonian relations between stresses and shear rates or velocity gradients, which may exhibit viscoelasticity and different constitutive relations (relations between density, pressure, and temperature) in comparison with Newtonian fluids. A non-Newtonian fluid may not have a well-defined viscosity. In those cases, the viscous stresses need to be represented differently. The following discussions and the subsequent examples of applications in this book are mainly restricted to Newtonian fluids.

The governing equations presented in Equations (1.1)–(1.5) have seven unknown flow-field variables: ρ, p, u, v, w, e, and T, but there are only five equations. In practical applications, there are always additional equations that can be used to close the system. For instance, in aerodynamics, it is generally reasonable to assume the gas is a perfect gas (which assumes that intermolecular forces are negligible). For a perfect gas, the equation of state is $p = \rho RT$, which provides a sixth equation, where R is the specific gas constant. A seventh equation to close the entire system is a thermodynamic relation between the state variables, for example, $e = e(T, p)$. For a calorically perfect gas (constant specific heats), this relation would be $e = c_v T$, where c_v is the specific heat at constant volume.

Using the definition of substantial derivative in Cartesian coordinates—$D/Dt \equiv \partial/\partial t + u\,\partial/\partial x + v\,\partial/\partial y + w\,\partial/\partial z$—the governing equations given in Equations (1.1)–(1.5) can also be written as

$$\frac{D\rho}{Dt} + \rho\nabla\cdot\mathbf{V} = 0 \tag{1.7}$$

$$\rho\frac{Du}{Dt} = -\frac{\partial p}{\partial x} + \frac{\partial\tau_{xx}}{\partial x} + \frac{\partial\tau_{yx}}{\partial y} + \frac{\partial\tau_{zx}}{\partial z} + \rho f_x \tag{1.8}$$

$$\rho\frac{Dv}{Dt} = -\frac{\partial p}{\partial y} + \frac{\partial\tau_{xy}}{\partial x} + \frac{\partial\tau_{yy}}{\partial y} + \frac{\partial\tau_{zy}}{\partial z} + \rho f_y \tag{1.9}$$

$$\rho\frac{Dw}{Dt} = -\frac{\partial p}{\partial z} + \frac{\partial\tau_{xz}}{\partial x} + \frac{\partial\tau_{yz}}{\partial y} + \frac{\partial\tau_{zz}}{\partial z} + \rho f_z \tag{1.10}$$

$$\rho\frac{D}{Dt}\left[\left(e+\frac{V^2}{2}\right)\right]$$

$$= \rho\dot{q} + \frac{\partial}{\partial x}\left(k\frac{\partial T}{\partial x}\right) + \frac{\partial}{\partial y}\left(k\frac{\partial T}{\partial y}\right) + \frac{\partial}{\partial z}\left(k\frac{\partial T}{\partial z}\right)$$

$$-\frac{\partial(up)}{\partial x} - \frac{\partial(vp)}{\partial y} - \frac{\partial(wp)}{\partial z} + \frac{\partial(u\tau_{xx})}{\partial x} + \frac{\partial(u\tau_{yx})}{\partial y} + \frac{\partial(u\tau_{zx})}{\partial z}$$

$$+\frac{\partial(v\tau_{xy})}{\partial x} + \frac{\partial(v\tau_{yy})}{\partial y} + \frac{\partial(v\tau_{zy})}{\partial z} + \frac{\partial(w\tau_{xz})}{\partial x} + \frac{\partial(w\tau_{yz})}{\partial y} + \frac{\partial(w\tau_{zz})}{\partial z} + \rho\mathbf{f}\cdot\mathbf{V}$$

$$\tag{1.11}$$

In CFD, Equations (1.7)–(1.11) are referred to as being of the nonconservation form, while Equations (1.1)–(1.5) are referred to as being of the conservation form or the divergence form. In general fluid dynamics, whether the equations are given in the conservation or nonconservation form is irrelevant. Indeed, through simple manipulations, one form can be obtained from the other. However, there are cases in CFD where one particular form is more important than the other (Anderson 1995). For instance, in flows containing shock waves, the computed flow field results are generally more smooth and stable when the conservation form of the governing equations is used. Also, the conservation form may provide a numerical and computing convenience.

The equations presented above are applicable to an unsteady, three-dimensional, compressible, viscous flow. Depending on the physical conditions, assumptions on the flow can be made for many cases, such as steady, one- or two-dimensional, incompressible, and inviscid flows. In CFD, governing equations under different assumptions are often employed to simplify the problem for different applications. For instance, inviscid flows are very important in aerodynamic applications. An inviscid flow is a flow where the dissipative, transport phenomena of viscosity, mass diffusion, and thermal conductivity are neglected. Taking the Navier–Stokes equations for a viscous flow and simply dropping all the terms involving friction and thermal conduction, the equations for an inviscid flow, the Euler equations, are obtained. The conservation form of the Euler equations can be written as follows.

Continuity equation:

$$\frac{\partial \rho}{\partial t} + \nabla \cdot (\rho \mathbf{V}) = 0 \tag{1.12}$$

Momentum equations (the Euler equations)

x component:

$$\frac{\partial (\rho u)}{\partial t} + \nabla \cdot (\rho u \mathbf{V}) = -\frac{\partial p}{\partial x} + \rho f_x \tag{1.13}$$

y component:

$$\frac{\partial (\rho v)}{\partial t} + \nabla \cdot (\rho v \mathbf{V}) = -\frac{\partial p}{\partial y} + \rho f_y \tag{1.14}$$

z component:

$$\frac{\partial (\rho w)}{\partial t} + \nabla \cdot (\rho w \mathbf{V}) = -\frac{\partial p}{\partial z} + \rho f_z \tag{1.15}$$

Energy equation:

$$\frac{\partial}{\partial t}\left[\rho\left(e + \frac{V^2}{2}\right)\right] + \nabla \cdot \left[\rho\left(e + \frac{V^2}{2}\right)\mathbf{V}\right]$$

$$= \rho \dot{q} - \frac{\partial (u p)}{\partial x} - \frac{\partial (v p)}{\partial y} - \frac{\partial (w p)}{\partial z} + \rho \mathbf{f} \cdot \mathbf{V} \tag{1.16}$$

Euler equations are governing equations for high-speed compressible flows where viscous effects may be neglected. They are important to many aerodynamic applications. Historically, Euler derived the continuity and momentum equations in the eighteenth century. He had little to do with the energy equation because thermodynamics as a science developed in the nineteenth century. In much of the aerodynamics literature, only the momentum Equations (1.13)–(1.15) are labeled as the Euler equations. However, in some CFD literature, the whole system of equations—continuity, momentum, and energy—are (inaccurately) referred to as the Euler equations. Although Euler equations are important equations for high-speed aerodynamics, they are of little relevance to DNS and LES because these equations neglect the dissipative effects, which are the most important characteristic of the small scales of turbulence.

Another important simplification of the fundamental governing equations (Navier–Stokes equations) is for incompressible flows, which will be discussed later in this chapter. The incompressible flow governing equations have significance in CFD, including DNS and LES.

B. Comments on the Governing Equations

The governing equations presented in Equations (1.1)–(1.5) or (1.7)–(1.11) are now broadly referred to as Navier–Stokes equations in honor of two men—the Frenchman M. Navier and the Englishman G. Stokes, who independently obtained the momentum equations in the first half of the nineteenth century. The terminology of Navier–Stokes equations was historically for the momentum equations only. However, in modern CFD literature, this terminology has been expanded to include the entire system of flow equations for the solution of a viscous flow—continuity and energy as well as momentum. This situation is similar to the Euler equations for inviscid flows. In addition, it needs to be noted that the equations presented above are for Cartesian coordinates. These equations can also be presented in cylindrical or spherical polar coordinates, or even in general curvilinear coordinates. Moreover, the equations discussed so far are for nonreacting single-phase flows. The governing equations for multicomponent reacting flow systems such as those encountered in combustion applications (e.g., Kuo 2005) and those for multiphase flow systems (e.g., Crowe 2006) can also be given, but they are more complex in form due to the involvement of multi-species and multiphases. It needs to be noted that in these complex flow systems, empirical approximations and/or correlations are also included in the governing equations in addition to the fundamental physical principles.

For the governing equations of fluid dynamics, it is observed that they are a coupled system of nonlinear partial differential equations, and hence are very difficult to solve analytically. To date, there is no general closed-form solution to these equations. (This does not mean that no general solution exists—it has just not been found so far.) In fact, this is one reason why CFD exists—numerical solutions of the governing equations have to be relied on rather than analytical solutions, which exist only for a number of very simplified cases such as two-dimensional boundary layer flows.

As Anderson (1995) discussed, the mathematical behavior of these partial differential equations has a significant impact on CFD. First, it is important that a CFD problem is well-posed, meaning that the solutions to the equations exist and are unique and the solutions depend continuously upon the initial and boundary conditions. In DNS and LES, boundary conditions are of crucial importance to all applications, whereas initial conditions are of only secondary importance to some special cases. Chapter 2 provides a discussion of various boundary conditions in detail. Second, it is important to classify whether the governing equations are hyperbolic, parabolic, or elliptic. Elliptic equations have to be solved simultaneously over the whole domain, whereas parabolic and hyperbolic equations are propagated from one location to another. The major mathematical behavior of hyperbolic and parabolic equations is that they lend themselves quite well to marching solutions. In contrast, for elliptic equations, the flow variables at a given point must be solved simultaneously with the flow variables at all other points. It can be shown that the unsteady Navier–Stokes equations have mixed behavior; they are in general elliptic in space and parabolic in time.

The fundamental governing equations apply to both laminar and turbulent flows. In principle, numerically solving these governing equations to a satisfactory degree of accuracy will provide the answers needed; this is indeed the concept of DNS. However, this is practically very difficult due to the huge amount of resources required for most of the practical problems, which are very often beyond the reach of present-day available computing power. For turbulence flows, the existence of a broad range of time and length scales will usually lead to prohibitively high computing costs if the equations were to be directly solved. The traditional CFD approach of turbulent flows has focused on the turbulent mean flow, in which the Reynolds- or ensemble-averaged governing equations are solved. In the literature, this approach is referred to as the RANS modeling of turbulence. In the RANS approach, it was shown that the time or ensemble averaging of the fundamental governing equations leads to the appearance of unknown correlations. For an incompressible flow (which will be discussed shortly), the unknown correlations are the Reynolds (or turbulent) stresses, $\rho \overline{u'_i u'_j}$ (transport of x_i momentum in x_j direction, primed variables represent fluctuating components of the variables), and turbulent heat fluxes, $\rho c_v \overline{u'_i T'}$. The averaged governing equations for turbulent flow, namely, the continuity, momentum, and energy equations, containing the unknown terms, therefore do not form a closed set. Determination of these unknown correlations is the subject of turbulence modeling. The complexity of turbulence makes it unlikely that any single model will be able to represent all turbulent flows. Thus RANS turbulence models should be regarded as engineering correlations or approximations rather than scientific laws. Experience with RANS-based turbulence models has yielded both successes and failures. This is why more advanced CFD such as DNS and LES has garnered attention in the past few decades.

Finally, it needs to be mentioned that it is always the same set of governing equations such as those shown in Equations (1.1)–(1.5) and (1.6)–(1.10) that govern the flow of a fluid, but the flow fields are quite different for different cases. Why? The answer is through the boundary conditions, which are quite different for each case. Therefore, the real driver for any particular flow field solution is the boundary conditions. This is also why boundary conditions are always very important in CFD, which will be discussed in Chapter 2 of this book.

C. Governing Equations for Incompressible Flows

The governing equations presented in Equations (1.1)–(1.5) and (1.6)–(1.10) are for compressible flows where fluid density changes significantly with the pressure. In CFD calculations of a compressible flow, the pressure, which is one of the unknowns, is directly calculated from an equation of state such as $p = \rho RT$ for a perfect gas. However, the fluid in many applications, such as liquid flows and gas flows at relatively low speed (usually with Mach number $M < 0.3$) where density is almost a constant, exhibits very little compressibility. For CFD analysis of such flows, the incompressible form of the governing equations is always used. In a practical CFD computation, there are severe problems if a compressible flow code is used for a nearly incompressible flow.

As the Mach number becomes smaller, compressible flow solvers suffer severe deficiencies in both efficiency and accuracy. The CFD solvers for compressible flows are normally density based, in which density is one of the variables in the solution while pressure is directly calculated from the density and temperature. The CFD solvers for incompressible flows, on the other hand, are normally pressure based, in which pressure is solved from a governing equation such as Poisson's equation for pressure.

The governing equations for incompressible flows can be conveniently obtained from those for compressible flows. The fundamental governing equations for an unsteady, three-dimensional, incompressible, and viscous flow can be given as follows:

Continuity equation,

$$\nabla \cdot \mathbf{V} = 0 \tag{1.17}$$

Momentum equations (the Navier–Stokes equations),

x component,

$$\rho \frac{Du}{Dt} = -\frac{\partial p}{\partial x} + \mu \nabla^2 u + \rho f_x \tag{1.18}$$

y component,

$$\rho \frac{Dv}{Dt} = -\frac{\partial p}{\partial y} + \mu \nabla^2 v + \rho f_y \tag{1.19}$$

z component,

$$\rho \frac{Dw}{\partial t} = -\frac{\partial p}{\partial z} + \mu \nabla^2 w + \rho f_z \tag{1.20}$$

Energy equation,

$$\rho \frac{De}{Dt} = \rho \dot{q} + \frac{\partial}{\partial x}\left(k\frac{\partial T}{\partial x}\right) + \frac{\partial}{\partial y}\left(k\frac{\partial T}{\partial y}\right) + \frac{\partial}{\partial z}\left(k\frac{\partial T}{\partial z}\right) + \rho \Phi + \rho \mathbf{f} \cdot \mathbf{V} \tag{1.21}$$

In the above equations, the Laplacian operator in Cartesian coordinates is defined as $\nabla^2 \equiv \partial^2/\partial x^2 + \partial^2/\partial y^2 + \partial^2/\partial z^2$. In obtaining these equations, the dynamic viscosity μ has been assumed to be a constant. For the energy equation, it has been assumed that the kinetic energy per unit mass $V^2/2 = (u^2 + v^2 + w^2)/2$ is much smaller than the internal energy and the work done by pressure is negligible. Thus, the energy equation is decoupled from the continuity and momentum equations.

In the energy equation, the rate of dissipation per unit mass is the work done by viscous stresses, given as

$$\Phi = v\left(\frac{\partial u_i}{\partial x_j} + \frac{\partial u_i}{\partial x_j}\right)\left(\frac{\partial u_i}{\partial x_j} + \frac{\partial u_i}{\partial x_j}\right) \tag{1.22}$$

where v is the kinematic viscosity.

For incompressible flows, the fundamental governing equations lack an independent equation for the pressure. The continuity equation cannot be used directly. The Poisson's equation for pressure, which can be derived from the continuity and momentum equations, is an important equation in CFD for incompressible flows. Taking the divergence of the momentum equations and then simplifying using the continuity equation, the Poisson's equation for pressure can be obtained as

$$\nabla^2 p = -\frac{\partial}{\partial x_i}\left[\frac{\partial(u_i u_j)}{\partial x_j}\right] \tag{1.23}$$

It can be shown that the Poisson's equation for pressure is an elliptic problem; that is, pressure values on boundaries must be known to compute the whole flow field. In CFD solvers for incompressible flows, solution of the Poisson's equation for pressure is an integral part of the code.

In fluid dynamics, the Boussinesq (1903) approximation is a concept that has been used in the field of buoyancy-driven flows and low-speed reacting flows. When such an approximation is used, density differences are neglected, except where they appear as the body forces (density differences multiplied by the gravitational acceleration **g**). When the Boussinesq approximation is used, the governing equations that must be solved are essentially those for incompressible flows, but the buoyancy effects due to the interaction between density inhomogeneity and gravity are taken into account.

In CFD, the concept of low Mach number governing equations is useful in many low-speed flow applications involving large density changes. The simplest low Mach number model is expressed by the incompressible

Navier–Stokes equations for a constant density fluid. Generalizations that incorporate variations in density include the Boussinesq approximation. Using asymptotic analysis, the set of governing equations for low Mach number flows can be derived from the original compressible Navier–Stokes equations (e.g., Müller 1998). From an analysis of the compressible Navier–Stokes equations, pressure in the flow field can be revealed as the zeroth-order global thermodynamic pressure, the first-order acoustic pressure, and the second-order "incompressible" pressure. The governing equations for low Mach number flows obtained using asymptotic analysis can be particularly useful for very weakly compressible flows encountered in combustion, aerodynamics, and aeroacoustics.

II. TURBULENCE AND DIRECT NUMERICAL SIMULATION

Most flows of engineering importance and flows that occur in our environment and in nature are turbulent. As one of the greatest challenges in science, turbulence has so far not been fully understood. Turbulent flows may be distinguished from laminar flows by their characteristics:

- Turbulent flows are by nature three-dimensional and unsteady and involve three-dimensional fluctuations.

- In turbulent flows, mixing of mass, momentum, or heat takes place far more effectively than by molecular diffusion in laminar flows.

- Turbulence has been viewed, conventionally, as a stochastic phenomenon. It is now established that most turbulent flow fields, such as boundary layers and free shear layers, exhibit a definite structure and some degree of order.

- Turbulence may be viewed as a vortical flow with a wide spectrum of eddy sizes and fluctuating frequencies; that is, it involves a wide spectrum of length and time scales.

- In turbulent flow, the large eddies are associated with low frequencies and small eddies with high frequencies. The large eddies are unstable and can break down into small eddies, which in turn break down into smaller eddies, and so on. There is, therefore, a continuous transfer of kinetic energy from the larger to the smaller eddies, a process that is referred to as the energy cascade. This continuous supply of energy, which is necessary to maintain the turbulence, is extracted from the main flow by the largest eddies and is finally dissipated into the smallest eddies.

- Turbulence is a continuum process; that is, the time and length scales of the smallest eddies are many orders of magnitude greater than the time scales and free paths of molecular motion.

- The dynamic and geometric characteristics of the large eddies are determined by the boundary conditions of the flow domain, and correspond closely with those of the mean flow.

- The largest eddies are therefore anisotropic. They are responsible for most of the turbulent mixing and contain a large portion of the total kinetic energy of turbulence.

- The smallest eddies, on the other hand, do not contribute significantly to the process of turbulent mixing and contain only an insignificant fraction of the total kinetic energy. The correspondence with the mean flow is absent. These eddies are therefore isotropic.

Irrelevant to the state of the flow (i.e., laminar or turbulent), the governing equations for fluid dynamics should apply because they are the mathematical statements of fundamental physical principles. The most exact approach to turbulence simulation is to solve the fundamental governing equations without any approximation or modeling. Direct numerical simulation is a CFD method that directly solves all the relevant time and length scales in the flow field. Since DNS captures all of the relevant scales of turbulent motion, no model (meaning simplification and approximation) is needed for the physical scales, including the small scales. For complex problems such as those encountered in most engineering applications, DNS is extremely expensive, if not intractable. The computational costs of DNS can be roughly estimated based on the analysis of the physical scales involved in a fluid flow, using the concept of Kolmogorov microscales.

In fluid dynamics, the understanding of turbulence scales was largely begun with the pioneering work of Kolmogorov (1941). Kolmogorov's theory is based on theoretical hypotheses combined with dimensional arguments and experimental observations. His contributions to turbulence have been reviewed numerous times (e.g., Hunt and Vassilicos 1991). Extensive descriptions on turbulence scales can be found in many texts (e.g., Hinze 1975, Pope 2000, Tennekes and Lumley 1972). Kolmogorov microscales are the smallest scales in turbulent flows. They are defined by

Kolmogorov length scale $\qquad \eta = \left(\dfrac{v^3}{\varepsilon} \right)^{1/4}$ $\hspace{3cm}$ (1.24)

Kolmogorov time scale $\qquad \tau_\eta = \left(\dfrac{v}{\varepsilon} \right)^{1/2}$ $\hspace{3cm}$ (1.25)

Kolmogorov velocity scale $\qquad u_\eta = (v\varepsilon)^{1/4}$ $\hspace{2.5cm}$ (1.26)

where ε is the average rate of energy dissipation per unit mass and v is the kinematic viscosity of the fluid. The Kolmogorov microscales are based on the idea that the smallest scales of turbulence are universal (similar for every turbulent flow) and that they depend only on ε and v. The definitions of the Kolmogorov microscales can be obtained by using this idea and dimensional analysis.

In a DNS, all the spatial scales, ranging from the smallest dissipative scales (Kolmogorov microscales) to the integral scale of the flow, which depends usually on the spatial scale of the boundary conditions, are associated with the motions containing most of the kinetic energy and need to be resolved directly in the computational mesh. Consider an integral scale L and number of discretized points N along a given mesh direction with uniform increments h, where the resolution has to satisfy $Nh > L$ so that the integral scale is contained within the computational domain. In the meantime, $h \leq \eta$ needs to be satisfied so that the Kolmogorov scale can be resolved. Since $\varepsilon \approx u'^3/L$, where u' is the root mean square (rms) of the velocity, the previous relations imply that a three-dimensional DNS requires a number of mesh points N^3 satisfying

$N^3 \geq \mathrm{Re}^{9/4} = \mathrm{Re}^{2.25}$, where $\mathrm{Re} = \dfrac{u'L}{v}$ is the turbulent Reynolds number

Therefore, the requirement for computing resources including CPU and memory storage requirement in a DNS grows very fast with the Reynolds number due to a significant increase in the required number of mesh points.

Since turbulence by nature is unsteady and three-dimensional, all DNS has to be time-dependent simulation. An intrinsic restriction on the time step Δt is that $\Delta t / \tau_\eta$ should be small. Also, from a numerical point of view, the integration of the solution in time has to be done by an explicit method because of the very large memory necessary. This means that in order to be accurate, the integration must be done with the time step Δt small enough such that a fluid particle moves only a fraction of the mesh spacing h in each step. Therefore, the time step has to be sufficiently small in order to be physically and numerically accurate and stable. Since the simulations need to be performed for a considerable period, such as the integral time scale, the costs on CPU are extremely high.

The combination of a large number of mesh points for spatial resolution and a small time step for time-marching leads to very high computational costs of DNS, even at low Reynolds numbers. For the Reynolds numbers encountered in most industrial applications, the computational resources required by DNS would exceed the capacity of the most powerful computers currently available. However, direct numerical simulation is a useful tool in fundamental research in turbulence. A well-defined DNS can be regarded as a detailed "numerical experiment," from which useful databases can be established. Analyzing the databases, information that is difficult or impossible to obtain in laboratory experiments can be extracted from the DNS results, allowing a better understanding of the physics of turbulence. Also, direct numerical simulations are useful in the development of turbulence models for practical applications, such as the models in RANS approach and the subgrid scale (SGS) models for LES. This is done by means of a priori tests, in which the input data for the model is taken from a DNS simulation, or by a posteriori tests, in which the results produced by the model are compared with those obtained by DNS.

DNS is a method in which all the scales of motion of a turbulent flow are computed. In other words, DNS directly resolves all the time and length scales of a fluid motion without modeling approximation. DNS can overcome the deficiencies of turbulence modeling. In the RANS approach, the complexity of turbulence makes it unlikely that any single turbulence model would be able to represent all turbulent flows. Turbulence models are essentially engineering correlations or approximations involving intrinsic inaccuracies. For RANS analysis of turbulent mean flow, a certain type of empirical data is required to determine the coefficients/constants in the turbulence models. Experience with RANS-based turbulence models has yielded both successes and failures. Similar to a RANS approach, LES is not model-free. The SGS models also involve coefficients/constants that need to be determined by empirical means. DNS can overcome these problems; it is only restricted to simple flows because of its high computational costs. DNS is now the method of choice for investigating the physics of turbulence for a number of simple flows such as shear layers. State-of-the-art DNS is mainly restricted to low Reynolds number turbulent flows in simple geometries. Current DNS can be classified into two categories:

- Temporal DNS: homogeneity in at least one direction

- Spatial DNS: inhomogeneity in all three directions

In temporal DNS, the turbulence is assumed to be homogeneous in at least one direction. This is an idealized situation that rarely approximates real flows. However, the assumption of homogeneity leads to mathematical simplicity. In the homogeneous directions, spectral methods that are highly accurate can be conveniently employed. Temporal DNS can be used to analyze mixing-layer or boundary-layer flows (e.g., Guo et al. 1995). However, they cannot deal with practical boundaries such as inflow or outflow boundaries, which are frequently encountered in CFD for engineering applications. Spatial DNS, on the other hand, is more suitable to practical engineering applications, but it has to deal with much higher computational costs. With the rapid increase in available computer powers, DNS has been predominantly performed as spatial DNS.

In practical DNS, there are several requirements that need to be satisfied so that all the relevant time and length scales can be adequately resolved:

- Time-dependent, three-dimensional simulations with mesh size smaller than the smallest physical length scales and time step smaller than the smallest physical time scales: to yield all of the flow variables at a large number of spatial locations for many instants of time.

- Highly accurate numerical schemes both in space and time with high-fidelity numerical boundary conditions.

The numerical method used in DNS has to be accurate enough. As in all CFD methods, numerical errors can lead to deterioration in the numerical solution. High-order numerical schemes are always required in DNS, even in LES. In a high-order scheme, the error in the functional approximation decreases much faster with the grid spacing than that in a lower-order scheme since the error is proportional to h^n for a numerical scheme with nth order formal accuracy, where h stands for the grid spacing. When accuracy is a crucial requirement of the simulation and a minimal truncation error has to be achieved, the application of high-order schemes allows the use of a coarser mesh compared to lower-order schemes, which may need excessively fine meshes to achieve the same accuracy. Although high-order schemes may require more calculations at individual points, such as involving more neighboring data points, this computational penalty can be effectively compensated for by the use of a coarser mesh. In a CFD simulation, the computing time depends mainly on the complexity of the method and the grid resolution. As a result of using a coarser mesh, one would expect a decrease in the computing time for a computationally efficient high-order scheme. In DNS and LES, the employment of high-order numerical schemes may significantly reduce the computing costs to achieve the required accuracy compared with using lower-order numerical schemes, providing that the high-order numerical schemes are not overly complex in formation and can be efficiently implemented. Strictly speaking, DNS is exact only if numerical schemes are accurate enough. Boundary conditions in DNS also play a vital role since they not only need to be able to represent the physical conditions as realistically as possible but also need to be compatible with the high-order numerical schemes.

DNS offers many attractive and advantageous features, which were summarized by Kasagi (1998) as

- DNS can be superior to experimental measurements in permitting full access to all the instantaneous flow variables, so that turbulent structures and transport mechanisms can be extensively analyzed.

- Experimental measurement techniques can be tested and evaluated against detailed and accurate DNS results.

- DNS can provide precise and detailed turbulence statistics, which is useful in evaluating and developing turbulence models.

- Effects of important parameters characterizing flow and scalar fields, such as Reynolds number, Prandtl number, and Schmidt number, can be systematically varied and examined.

- DNS can offer an opportunity to accurately study a virtual flow field that would not occur in reality.

Historically, the meteorologist Richardson (1922) proposed numerical schemes to solve the equations of fluid mechanics applied to the atmosphere in a deterministic fashion; this marked the beginning of CFD and DNS as the simplest CFD in its mathematical formulation of the governing equations—nothing more than the original Navier–Stokes equations (but most complex in numerical methods and boundary conditions). The terminology of DNS, however, is widely accepted as being built upon the foundation work at the U.S. National Center for Atmospheric Research in 1972 by Orszag and Patterson (1972). Certainly DNS evolves with the development of numerical schemes and computer hardware. In recent years, a few very big DNS have been performed in Japan. DNS using 4096^3 mesh points was carried out in the Japanese Earth Simulator supercomputer (Sato 2004) in 2002, which still represents one of the biggest DNS performed in the world nowadays.

The applications of DNS so far have mainly focused on shear layer flows. A shear layer is a relatively narrow region within the fluid flow where a rapid variation in velocity normal to the direction of the velocity takes place. Examples of shear layers consist mainly of a boundary layer on a flat plate, flow in pipes and channels, jets and plumes, and wakes. Although shear layer flows are simpler than many flows encountered in practical engineering applications, shear layer flows are "building blocks" for many more complex flows. They also provide ideal cases for model development. Some examples are shown for applications of DNS to these shear layer flows in subsequent chapters.

III. LARGE-EDDY SIMULATION

Large-eddy simulation was also based on Kolmogorov's (1941) famous theory on turbulence. The theory makes the assumption that large eddies of the flow are dependent on the flow geometry, whereas smaller eddies are self-similar and have a universal character. It becomes a practice to solve only for the large eddies explicitly, and model the effect of the smaller and more universal eddies on the larger ones. In LES, the large-scale motions of the flow are calculated similar to those in DNS, whereas the effect of the smaller universal scales (the so-called subgrid scales) are modeled using a subgrid scale (SGS) model. In practical implementations, one is required to solve the filtered Navier–Stokes equations with additional SGS stress terms. One can think of the method as applying DNS to the large scales and modeling to the small scales as in the RANS approach.

There are differences between the modeling in LES and RANS. In RANS, the modeling is based on the time- or ensemble-averaged governing equations; therefore, it would not be able to capture accurately the flow unsteadiness and the dynamics of small scales, in a sense that the average of a fluctuating quantity is taken as zero, such as $\overline{u'} = 0$. In LES, the governing equations are spatially filtered rather than ensemble or time averaged. Explicit account is taken of flow structures larger than the filter width, whereas the influence of unresolved scales is modeled using an SGS model. The justification for LES is that the larger eddies contain most of the energy, do most of the transporting of conserved properties, and vary most from flow to flow; the smaller eddies are believed to be more universal and less important and should be easier to model. It is hoped that universality is more readily achieved at this level than in RANS modeling but this assertion remains to be proved. Unlike RANS, filtering is used rather than averaging; more important, the small-scale component of a quantity is no longer zero, such as $\overline{u'} \neq 0$. CFD practice has shown that LES can significantly improve predictions of vortical and other complex unsteady flow structures in the flow fields, which RANS very often fails to do.

In LES, it is essential to define the quantities to be computed precisely as in the RANS case. To do this it is crucial to define a velocity field that contains only the large-scale components of the total field. This is best done by filtering; the large or resolved scale field is essentially a local average of the complete field. For one-dimensional flow, the filtered velocity is defined by

$$\overline{u}_i = \int G(x,x')u_i(x')dx' \tag{1.27}$$

where $G(x, x')$, the filter kernel, is a localized function or a function with compact support, that is, one that is large only when x and x' are not far apart. Filter kernels, which have been applied in LES, mainly include Gaussian, box, and cutoff.

When the Navier–Stokes equations are filtered, one obtains a set of equations very similar in form to the RANS equations. For an incompressible flow without body force, the filtered Navier–Stokes equations used in LES is given as

$$\frac{\partial \overline{u}_i}{\partial t} + \frac{\partial \overline{u_i u_j}}{\partial x_j} = -\frac{1}{\rho}\frac{\partial \overline{p}}{\partial x_i} + \nu\frac{\partial^2 \overline{u}_i}{\partial x_j \partial x_j} \tag{1.28}$$

where the definitions of the velocities differ from those in RANS, but the closure issues are very similar. Since $\overline{u_i u_j} \neq \overline{u}_i \overline{u}_j$, a modeling approximation for the difference between the two sides of this inequality $\tau_{ij} = \overline{u_i u_j} - \overline{u}_i \overline{u}_j$ must be introduced. In the context of LES, $\tau_{ij} = \overline{u_i u_j} - \overline{u}_i \overline{u}_j$ is called the subgrid scale (SGS) Reynolds stress. It plays a role in LES similar to the role played by the Reynolds stress in RANS models but the physics that it models is different. By writing the complete velocity field as a combination of the filtered field and a subgrid scale field, we can decompose the subgrid scale Reynolds stress (SGSRS) into three sets of terms:

$$\tau_{ij} = (\overline{\overline{u}_i \overline{u}_j} - \overline{u}_i \overline{u}_j) + (\overline{\overline{u}_i u'_j} + \overline{\overline{u}_j u'_i}) + \overline{u'_i u'_j} \tag{1.29}$$

which may be ascribed with physical significance. In particular, these three terms represent the following physics (Ferziger 1996):

- The first term represents the interaction of two resolved scale eddies to produce small-scale turbulence. It has been called the Leonard term and, sometimes, the outscatter term.

- The second term represents the interaction between the resolved scale eddies and the small-scale eddies. This term, also called the cross term, can transfer energy in either direction but, on average, transfers energy from the large scales to the small ones.

- The third term represents the interaction between two small-scale eddies to produce a large-scale eddy and is called the true subgrid scale term. It is also called the backscatter term.

The SGS Reynolds stress in LES and the Reynolds stress in RANS are physically and numerically different. The SGS Reynolds stress in LES is due to a local average of the complete field, whereas the Reynolds stress in RANS is due to a time or ensemble average. The SGS energy is a much smaller part of the total flow than the RANS turbulent energy, so model accuracy may be less crucial in LES than in RANS computations. Subgrid scale modeling is the most distinctive feature of LES.

By far the most commonly used subgrid scale model is the one proposed by Smagorinsky (1963), which marked the beginning of LES. It is an eddy viscosity model, given as

$$\tau_{ij} - \frac{1}{3} \tau_{kk} \delta_{ij} = -\nu_T \left(\frac{\partial \overline{u}_i}{\partial x_j} + \frac{\partial \overline{u}_j}{\partial x_i} \right) = -2\nu_T \overline{S}_{ij} \tag{1.30}$$

where $\bar{S}_{ij}$ is the resolved strain rate tensor, δ_{ij} is the Kronecker delta, and v_T is the eddy viscosity. Although the Smagorinsky model was initially developed for atmospheric or oceanic flows, it was not a success for the predictions of atmospheric or oceanic dynamics because it overly dissipates the large scales (Lesieur et al. 2005).

Over the last two decades, there were many other SGS models of turbulence proposed. For instance, a self-consistent dynamic approach to evaluate the coefficients that appear in the subgrid model was developed by Germano et al. (1991). The dynamic approach developed by Germano et al. (1991) for incompressible flows was successfully extended to compressible flows by Moin et al. (1991). Erlebacher et al. (1992) extended the LES subgrid models for compressible flows. Nevertheless, the majority of LES study has been performed for incompressible flows and there are a few texts on LES for incompressible flows (e.g., Sagaut 2006). LES of compressible flows is relatively scarce. For a compressible flow, the SGS Reynolds stresses are more complex than those for an incompressible flow. The details are discussed in subsequent chapters. Furthermore, SGS models for complex flows such as those for multiphase flows and reacting flows encountered in combustion applications still remain to be developed and validated. For combustion applications, chemical reaction and the associated heat release introduce fine-scale density and velocity fluctuations that, in turn, couple the small-scale events back to the larger fluid-dynamic scales. Menon et al. (1993) applied the linear eddy mixing (LEM) model to LES of combustion. Within the context of LES, the LEM approach can be used to model the small-scale processes ranging from the grid resolution down to the Kolmogorov scale or the smallest scales related to chemical reaction in reduced dimension, whereas the large scales of the flow are calculated directly from LES equations of the motion with an appropriate coupling procedure. In the meantime, SGS models for complex multiphase flows are very immature. There is a lack of well-established SGS models, especially for the interactions between the different phases. There is no SGS model available to date that can take into account the subgrid influence of one phase that is locally smaller than the grid size (for instance, fine liquid droplets or solid particles dispersed in a gas medium) on the resolved scales.

There are similarities and differences between DNS and LES. Both DNS and LES have to be time-dependent three-dimensional simulations, due to the nature of turbulence. Both of them require high-order numerical schemes. In state-of-the-art DNS, the discretization schemes used are at least fourth order, typically sixth and above. In LES, the numerical schemes used are normally between second and fourth order. LES has been successfully applied to many industrial problems, but DNS has been mainly restricted to simple physical problems to understand the flow physics. LES can be applied to complex geometry problems, but the current DNS can deal with only simple geometries. DNS offers the most accurate flow predictions without modeling uncertainty. In terms of computational costs, however, DNS is much more expensive than LES.

Compared with traditional CFD based on the RANS approach, the main advantage of LES approaches is the increased level of detail it can deliver. While RANS methods provide "averaged" results, LES is able to predict instantaneous flow characteristics and resolve turbulent flow structures. This is important in many applications. For instance, in a combustion application, the "averaged" concentration of chemical species and temperature from RANS may be too low to trigger a chemical reaction; "instantaneous data" from LES, however, can be used to predict cases of localized areas of high concentration and temperature in which reactions do occur. For near-wall heat transfers, the instantaneous wall heat flux can be much higher or lower than the "averaged" value, which can be important to certain applications. The main advantage of RANS over LES is that RANS can be much cheaper than LES in terms of computational costs due to a possible coarser mesh and possible steady-state flow simulations in RANS (which is often the case). In addition, RANS can be used to study two-dimensional flows.

In general, LES has reduced modeling impact compared to RANS. LES offers significantly more accurate results than RANS for flows involving vortical structures and separation and for acoustic predictions. Over the last two decades, the hybridization of LES and RANS has drawn much attention in CFD, mainly for wall-bounded flow problems. Most SGS models display an inability of correctly accounting for the anisotropy and disequilibrium in near-wall regions. For LES, the computational demands increase significantly in the vicinity of walls if the near-wall flow motions are going to be directly resolved, and simulating such flows usually exceeds the limits of available computers. For this reason, zonal approaches are often adopted, with RANS or other empirically based models replacing LES in the wall region. Hybrid LES/RNS approaches as such can be very useful in many practical applications before robust near-wall models are developed for LES.

As advanced CFD tools, in both DNS and LES, boundary conditions are an integral and important part of the numerical solution of the governing equations. Boundary conditions for DNS and LES have their special features compared with those for RANS-based CFD. In addition, both DNS and LES have to be time-dependent simulations, in contrast to RANS-based CFD, which are very often steady-state simulations. In the following two chapters, numerical treatment of boundary conditions and discrete time integration methods for DNS and LES are discussed, and different applications of DNS and LES are discussed in subsequent chapters.

REFERENCES

Anderson, J.D. 1995. *Computational fluid dynamics: The basics with applications.* New York: McGraw-Hill.

Boussinesq, J.V. 1903. *Theorie analytique de la chaleur,* vol. 2. Paris: Gauthier-Villars.

Crowe, C.T. (editor) 2006. *Multiphase flow handbook.* Boca Raton, FL: Taylor & Francis Group.

Erlebacher, G., Hussaini, M., Speziale, C., and Zang, T. 1992. Toward the large-eddy simulation of compressible turbulent flows. *Journal of Fluid Mechanics* 238: 155–185.

Ferziger, J.H. 1996. Chapter 3 Large eddy simulation. In *Simulation and Modeling of Turbulent Flows,* ed. T.B. Gatski, M.Y. Hussaini, and J.L. Lumley, 109–154. Oxford, UK: Oxford University Press.

Germano, M., Piomelli, U., Moin, P., and Cabot, W. H. 1991. A dynamic sub-grid scale eddy viscosity model. *Phys. Fluids* A 3: 1760–1765.

Guo, Y., Adams, N.A., and Kleiser, L. 1995. Modeling of nonparallel effects in temporal direct numerical simulations of compressible boundary-layer transition. *Theoretical and Computational Fluid Dynamics* 7: 141–157.

Hinze, J.O. 1975. *Turbulence.* New York: McGraw-Hill.

Hunt, J.C.R. and Vassilicos, J.C. 1991. Kolmogorov's contribution to the physical and geometrical understanding of turbulent flows and recent developments. *Proc. R. Soc. Lond.* A 434: 183–210.

Kasagi, N. 1998. Progress in direct numerical simulation of turbulent transport and its control. *International Journal of Heat and Fluid Flow* 19: 125–134.

Kolmogorov A.N. 1941. Local structure of turbulence in an incompressible liquid for very large Reynolds numbers. *Dokl. Akad. Nauk SSSR* 30: 301–305.

Kuo, Kenneth K. 2005. *Principles of combustion.* Hoboken, NJ: John Wiley & Sons.

Lesieur, M., Métais, O., and Comte, P. 2005. *Large-eddy simulations of turbulence.* New York: Cambridge University Press.

Menon, S., McMurtry, P.A., and Kerstein, A.R. 1993. A linear eddy flamelet sub-grid model for large-eddy simulations of turbulent premixed combustion. In *Large eddy simulations of complex engineering and geophysical flows,* ed. B. Galperin, 288–314. Cambridge, UK: Cambridge University Press.

Moin, P., Squires, K., Cabot, W., and Lee, S. 1991. A dynamic sub-grid scale model for compressible turbulence and scalar transport. *Physics of Fluids* A 3: 2746–2757.

Müller, B. 1998. Low-Mach-number asymptotics of the Navier–Stokes equations. *Journal of Engineering Mathematics* 34: 97–109.

Orszag, S. A. and Patterson, G.S. 1972. Numerical simulation of three-dimensional homogeneous isotropic turbulence. *Physics Review Letters* 28: 76–69.

Pope, S.B. 2000. *Turbulent flows.* Cambridge UK: Cambridge University Press.

Richardson, Lewis F. 1922. *Weather prediction by numerical process.* Cambridge: Cambridge University Press.

Sato, T. 2004. The earth simulator: Roles and impacts. *Nuclear Physics B Proceedings Supplements* 129: 102–108.

Sagaut, P. 2006. *Large Eddy Simulation for incompressible flows: An introduction.* Berlin: Springer.

Smagorinsky, J. 1963. General circulation experiments with the primitive equations. *Monthly Weather Review* 91: 99–164.

Tennekes, H. and Lumley, J.L. 1972. *A first course in turbulence.* Cambridge, UK: MA: MIT Press.

CHAPTER 2

Numerical Treatment of Boundary Conditions

THE GOVERNING EQUATIONS SHOWN in Chapter 1 are a coupled system of nonlinear partial differential equations, and their numerical solutions depend on boundary and initial conditions. As discussed in Chapter 1, the Navier–Stokes equations are in general elliptic in space and parabolic in time. To solve these equations for a specific flow configuration under consideration, boundary conditions (BC) at all boundaries of the computational domain and initial conditions for all flow variables in the entire field are required. There are situations where initial conditions are important, such as predictions of the transition process or fundamental investigations of turbulence. An example is the decay of a homogeneous isotropic turbulent flow (Hinze 1975) often used for basic investigations in turbulence research. However, in most applications of DNS and LES, the initial conditions play a subsidiary role because the statistically steady-state flow status is of major concern, which should be reached independently from the initial conditions. In many cases, appropriate initial conditions can be chosen to shorten the simulation time until a statistically steady state is achieved.

Boundary conditions play a significant role in all numerical simulations of fluid flow, heat transfer, and combustion problems. They are the real driver for a particular CFD problem—the flow fields are quite different for different cases although the governing equations are the same. The differences are all due to the different boundary conditions for different problems, as mentioned in Chapter 1. The specification of boundary conditions is an important issue in all CFD calculations, which also needs special care in DNS or LES mainly for two reasons: (1) boundary conditions in DNS and LES have to represent the nature of the turbulent flow at the boundary, notably the unsteadiness; and (2) boundary conditions in DNS and LES need to be compatible with the high-order numerical schemes used for discretization. There is usually a significant difference between BC for RANS-based CFD and those for DNS or LES, while in principle no fundamental difference exists between DNS and LES concerning the challenge of appropriate BC.

In all CFD calculations, whether the traditional RANS modeling approach or the more advanced DNS or LES, boundary conditions must represent the physical conditions at the boundaries as faithfully as possible. According to the mathematical characteristics of differential equations, boundary conditions may be categorized as follows:

- Dirichlet BC: The value of the flow variable is specified. Dirichlet boundary conditions are typically associated with problems involving inflow phenomena and isothermal walls.

- Neumann BC: The normal gradient of the flow variable is specified. Neumann boundary conditions are often associated with symmetry boundaries and adiabatic walls.

- Mixed BC: A combination of Dirichlet and Neumann type BC is specified.

Boundary conditions can also be distinguished physically (Breuer 2007) as physical boundaries, such as a solid wall or artificial boundaries requiring physically meaningful approximations of the flow such as an outflow boundary. Artificial boundaries appear if the computational domain constitutes only a part of the total flow field, which is always necessary when the region of major interest is investigated instead of the entire flow field in order to reduce the computational costs. Artificial boundary conditions require physically meaningful approximations of the flow and are often difficult to formulate.

In the following sections, the most important types of boundary conditions for DNS and LES are discussed, including inflow and outflow boundaries, wall boundaries, and other boundaries such as far field and open boundaries, and periodic and symmetry boundaries. Particular attention has been given to the Navier–Stokes characteristic boundary conditions (NSCBC) for compressible flows by Poinsot and Lele (1992), which have been widely used in DNS and LES. Boundary conditions for incompressible flows are also discussed.

I. INFLOW AND OUTFLOW BOUNDARY CONDITIONS

Inflow and outflow boundary conditions are very important BC widely used in CFD because they represent the starting and ending conditions of the flow in its streamwise direction. Outflow BC always represent an artificial cut through the flow field at a downstream location, which is also true for the inflow at an upstream location. The difference between an inflow and an outflow BC is that there is always certain known information of the inflow for a practical application such as those from an upstream nozzle. It is more difficult to have known information on an outflow boundary where the flow is going out of the computational domain and may still be developing further downstream. In the following sections, a general discussion is given on inflow and outflow BC for DNS and LES, followed by discussions on inflow/outflow BC formulation based on the NSCBC for compressible flows. Finally, inflow/outflow BC for incompressible flows are discussed.

A. Inflow BC for DNS and LES

Inflow boundary conditions for DNS and LES are significantly different from those used in a RANS modeling approach. In a RANS approach, the inflow BC is normally of Dirichlet type with prescribed values that do not usually change with time. Inflow BC typically represent an artificial cut through the flow field. The inflow BC for DNS and LES has to be time dependent. DNS and LES computations require appropriate inflow data of Dirichlet type that adequately represents the flow field at an upstream location where the inflow locates. Meanwhile, the inflow data needs to be time dependent, to reflect the unsteadiness of vortices and turbulence. In a DNS or LES computation, the generation of artificial inflow data can be based on the knowledge of the flow geometry and/or some experimental data. In many cases, the instantaneous velocity u_i at the inflow may be split into two parts (according to Reynolds' approach): a steady, constant mean value $\bar{u}_i$ and a fluctuating component u_i'. Typically the mean value is known from experiments such as measured velocity profiles for nozzle flow, channel flow, and so on, or theory such as the simplified swirling velocity profile obtained from theoretical analysis (Jiang et al. 2008). Consequently, the generation of artificial inflow data can be restricted to the fluctuating component. There are a variety of techniques that can be used.

- Random number or white noise: Random numbers can be easily generated by a computer program such as the random number generator in FORTRAN. This random number method is apparently the simplest technique, but unfortunately also the worst technique because it very often does not work. Computer-generated random data is normally of a very high frequency, changing rapidly and randomly between time steps. There is no way to take any spatial or temporal correlation into account. However, the velocity field of a real turbulent flow does have certain correlations. The high-frequency random data has nothing in common with the physical situation. Furthermore, the unsteadiness in the high-frequency random data can be easily damped out by the numerical methods used to solve the governing equations. Therefore, the results obtained from using these inflow conditions are more or less identical to those obtained using a constant laminar inflow. For an inflow BC using a random number generator, the only free parameter to adjust is the root mean square (rms) value of the fluctuations.

- Stochastic fluctuations with a prescribed energy spectrum: Inflow BC of this type tries to provide a more realistic turbulent inflow by taking low wave numbers or low frequencies in a turbulent flow field into account. There have been a number of approaches for BC of this type. In general, the idea of all methods proposed is to produce inflow data that satisfies certain statistical properties such as rms values, cross-correlations, higher-order moments, length and time scales, or energy spectra. Klein et al. (2003) provided an overview of these techniques, which typically involve two steps: (1) generation of a provisional three-dimensional signal for the velocity components, and (2) cross-correlations between different velocity components using the method proposed by Lund et al. (1998). In the first step, a provisional three-dimensional signal for each velocity component is generated that possesses a prescribed two-point statistic. This can be achieved by an inverse Fourier transform, as described by Lee et al. (1992). It can also be achieved by digital filtering of the random data (Klein et al. 2003), which allows a prescribed second-order (one-point) statistic as well as autocorrelation functions to be reproduced. Di Mare et al. (2006) also discussed the method of using a digital filter reproducing specified statistical data. Compared with the inverse Fourier transform method, the digital filter method offers several advantages such as simplicity, flexibility, and accuracy. The method of using the stochastic fluctuations with a prescribed energy spectrum provides a useful means of specifying inflow BC for DNS and LES of turbulent flows. However, individual inflows encountered in practical applications may not always follow a prescribed energy spectrum.

- Synthetic eddy method and proper orthogonal decomposition for inflow data generation: These are efforts similar to the method of using the stochastic fluctuations with a prescribed energy spectrum. The idea behind the synthetic eddy method (Jarrin et al. 2006) is to focus directly on prescribing coherent structures in the inflow. It tries to reproduce prescribed first- and second-order one-point statistics, characteristic length and time scales, and the shape of coherent structures. The final velocity field is reconstructed from the mean velocity profile and the "vortex" velocity field. Druault et al. (2004) used a proper orthogonal decomposition (POD) and linear stochastic estimation to construct inflow conditions, which provided a useful estimation of the large-scale coherent structures at the inflow. However, this POD-based technique cannot be applied systematically for general flows as it requires a previous realization of the flow.

- Perturbed laminar inflow: Inflow BC based on a simple perturbation added to a mean velocity profile has been commonly used in DNS and LES. There are situations where the flow is laminar at the inflow and the transition to turbulence takes place at downstream locations within the computational domain as a part of the solution. For these applications, a simple velocity perturbation may be used and the frequency of the perturbation can be obtained from a linear stability analysis of the parallel base flow or from experiments. For instance, there was experimental evidence on the most unstable frequency leading to the jet preferred mode of instability (Hussain and Zaman, 1981), which can be used to trigger the formation of vortical structures in DNS and LES of jet flows. In many cases, a small perturbation of the sinusoidal type involving one or a few frequencies (the fundamental frequency and its harmonics) may be added onto the mean velocity profile, which will subsequently lead to the development of flow instabilities and the formation of vortical structures in the flow field (e.g., Jiang et al. 2007a; 2007b). If the flow is of a high enough Reynolds number, the large vortical structures will break down into smaller ones with transition to turbulence taking place at downstream locations. The type of perturbations used, especially the perturbing frequencies, will have a significant impact on the flow. For instance, vortex paring or merging (Mitchell et al. 1999; Sandham 1994) can take place in the flow field at downstream locations. In other cases, helical disturbances will lead to jet flapping modes (e.g., Danaila and Boersma 2000; Uchiyama 2004). In general, the rms or the magnitude of the perturbations does not affect the flow too much if it is not too large. (In many simulations the perturbation amplitude used is only around 1%.) When the perturbation is very large, the flow will be of a pulsating nature (Jiang et al. 2004; Jiang et al. 2006).

- Inflow from an auxiliary simulation: Obtaining inflow data for DNS or LES from an auxiliary or precursor simulation is an accurate technique to provide inflow boundary conditions. In this way, the inflow represents exactly the flow at the domain inlet. An example of this is the flow through a 90° bend (Breuer 2007). To generate appropriate inflow conditions for this case, an additional simulation for a straight duct with the same cross sections has to be carried out with periodic boundary conditions in the homogeneous flow direction. The instantaneous data from one plane of the auxiliary simulation can then be applied as the inflow boundary conditions for the inhomogeneous flows. However, this technique has two major drawbacks. First, the method lacks generality and is restricted to simple cases where the flow at the inlet of the computational domain can be regarded as the solution of another flow. Second, the method entails a heavy extra computational load and may require large storage capacities. An auxiliary simulation will certainly cost computing resources, which can be a significant amount if the auxiliary simulation has to be performed for a large domain using lots of grid points for many time steps. In addition, if the auxiliary simulation and the main simulation are not running in parallel, an enormous amount of data will have to be stored.

- Other inflow BC—periodic BC and inflow BC without perturbations: The use of periodic BC will avoid inflow and outflow completely, and has been used in DNS and LES (e.g., temporal DNS). However, the applicability of periodic BC is restricted to flow configurations that are indeed periodic owing to their geometry, such as flow around turbine blades, or channel, pipe, and duct flows with one or more statistically homogeneous flow directions. Spalart and

Leonard (1987) and Spalart (1988) extended the application of periodic BC to turbulent boundary layers by using a coordinate transformation. However, there is a lack of generality for the application of periodic BC. In DNS and LES, there are also situations where it is adequate to use an inflow BC based on the mean velocity profile only, without including any external perturbations or excitations. For instance, simulations of buoyancy-driven jet and plume flows exhibit natural instabilities without external perturbations (Jiang and Luo 2000a; Jiang and Luo 2003), where the flow develops into vortical structures and turbulence at downstream locations on its own. In these simulations, the initial perturbation required to trigger the flow instabilities might have been provided by the numerical disturbance arising from the mismatch between the initial condition and the solution of Navier–Stokes equations. Once the instabilities are triggered in the initial stage of the simulation, they are self-sustainable. Under these circumstances, perturbations, especially spatial perturbations, may still be needed to break the symmetry in the flow field if necessary, but temporal perturbations will be unnecessary to the development of vortical structures in the flow field.

An appropriate BC, depending on the type of application, would be required from the above list of inflow boundary conditions. Apart from the method based on random number perturbation or white noise, all the other approaches can be effective and physically meaningful. In the above discussion, focus has been given to velocities, which comprise a very important flow property defining the downstream flows. The specification of other flow variables such as density and temperature at the inflow follow the same principle. Above all, an inflow boundary should be able to represent the real flow at the domain inlet as faithfully as possible. Experimental data can certainly be used as the inflow condition when it is available. Physically, it is often preferable to have fixed values of all the variables (plus their fluctuations when applicable) for an inflow BC. However, a prescribed inflow BC does not normally satisfy the solution of the governing equations in the vicinity of the inflow. Mathematically speaking, such an inflow with all the variables fixed does not meet the requirements for well posedness of the Navier–Stokes equations. When modern nondissipative algorithms are used, numerical oscillations occur in the vicinity of the inflow boundary if all the variables are fixed. Poinsot and Lele (1992) proposed the

Navier–Stokes characteristic boundary condition (NSCBC) for DNS and LES, which is a general formulation method for boundary conditions and is based on the analysis of characteristics. The NSCBC associated with the high-order nondissipative numerical algorithms uses the correct number of boundary conditions required for well-posedness of Navier–Stokes equations that can avoid numerical instabilities and spurious wave reflections at the computational boundaries. In NSCBC, the local one-dimensional inviscid (LODI) relations are used to provide compatible relations between the physical boundary conditions and the amplitudes of characteristic waves crossing the boundary. NSCBC can be used to decide the "numerical" or "soft" boundary condition at the inflow in a DNS or LES calculation, while other physical conditions at the inflow are specified or fixed with their known values (mean value or mean value plus fluctuating value using one of the methods discussed above). NSCBC has been broadly used in DNS and LES, which will be discussed in detail subsequently.

B. Outflow BC for DNS and LES

An outflow boundary represents an artificial cut through the flow field, similar to an inflow boundary. An outflow BC is used when a finite domain has to be adopted in order to avoid a prohibitively large domain for the entire flow field or when the downstream flow is no longer of interest. The major difference between an inflow boundary and an outflow boundary is that there is no information available on the flow outside the computational domain for an outflow, whereas such information is always available from the upstream conditions for an inflow. The flow variables at an outflow have to be approximated in a physically meaningful manner in order not to influence the solution of the governing equations within the computational domain. For an outflow BC, the upstream-traveling numerical reflection or perturbation triggered by the artificial outflow BC has to be eliminated or minimized. For the traditional steady-state RANS-based CFD, the condition of a fully developed flow in the main flow or streamwise direction is most often used for the outflow BC; that is, $\partial \phi / \partial x_i = 0$ with ϕ representing any flow variables. This is essentially a zero-order polynomial extrapolation. There can also be first- or second-order polynomial extrapolation. However, for DNS and LES, these boundary conditions are highly reflective, in that the solution in the vicinity of the outflow BC will be distorted or polluted by the unknown flow field outside the computational domain. This is not surprising because the accuracies of these conditions are not physically assured. DNS or LES predictions are inherently unsteady and dominated by dynamic vortical structures, and a simple zero-order extrapolation is not able to represent the flow unsteadiness and dynamic vortical structures.

Appropriate outflow boundary conditions for DNS and LES have to ensure that vortices can approach and pass the outflow boundary without significant disturbance or reflection back into the computational domain. For this purpose, an outflow BC referred to as the convective boundary condition has been widely used and is given as

$$\frac{\partial \phi}{\partial t} + U_{conv} \left.\frac{\partial \phi}{\partial x_i}\right|_{outflow} = 0 \tag{2.1}$$

Equation (2.1) represents nothing other than a simplified and linearized one-dimensional transport equation in the main flow direction x_i, where U_{conv} denotes a mean convective velocity in the x_i direction and has to be adjusted with respect to the flow simulated. One criterion for the appropriate choice of U_{conv} may be the mass flow rate at the outlet. For incompressible flows, this mass flow rate has to balance the inflow mass flux or satisfy the global mass conservation. In this case, U_{conv} is chosen to be a constant. For compressible flow, U_{conv} may be decided by averaging the flow results over a region close to the outflow boundary and continuously updated in the simulation. The spatial gradient in Equation (2.1) is approximated by one-sided differences. Breuer (2007) has shown that this convective outflow boundary condition can avoid the propagation of errors from the outflow boundary into the computational domain. The convective outflow boundary condition has been successfully used in both internal and external flows.

The difficulty of an outflow boundary condition is always associated with the upstream-traveling numerical reflection or perturbation triggered by the artificial outflow BC, which leads to propagation of errors from the outflow boundary into the computational domain. In order to minimize this reflection, a variety of so-called nonreflecting boundary conditions have been developed (e.g., Thompson 1987), especially for compressible flows. However, nonreflecting BC alone may not be sufficient in controlling the reflection from an outflow BC in a DNS or LES calculation because the flow is not necessarily going out of the domain instantaneously in the direction normal to the outflow boundary due to the existence of multidimensional vortical structures. To overcome this problem, a "sponge layer" may be used (e.g., Jiang et al. 2004). A sponge layer next to the outflow boundary can be used to control the spurious wave reflections from the outside of the computational domain, by manipulating the flow in the sponge layer so that it approaches the outflow boundary at a normal angle and the flow is instantaneously going out of the domain at the outlet. This can be achieved by changing the solution in the sponge layer to approach the averaged values of the flow variables obtained from an upstream "averaged zone" where the results are averaged over a region next to the sponge layer. The idea of using a sponge layer at the outlet of the domain is similar to that of the "sponge region" or "exit zone" (Mitchell et al. 1999), which has been proved to be very effective to control the wave reflections through the outflow boundary. The results in the sponge layer are not truly physical and therefore should not be used in the data analysis.

C. Inflow/Outflow BC Based on NSCBC for Compressible Flows

The Navier–Stokes characteristic boundary condition (NSCBC) proposed by Poinsot and Lele (1992) has been broadly used in the DNS and LES of compressible flows. The NSCBC can be used to decide the "numerical" or "soft" boundary condition at the inflow in DNS or LES, where the "soft" variable is being calculated during the simulation according to the characteristic waves across the boundary rather than fixed, and other variables are specified or fixed with their known values. The nonreflecting characteristic boundary condition (e.g., Thompson 1987), which has been used broadly as an outflow BC for compressible flows, can also be regarded as a special case of NSCBC due to the fact that both NSCBC and nonreflecting BC are both based on the analysis of characteristics of compressible flows. The notable difference between the NSCBC and the nonreflecting BC is that there is normally only one "soft" variable in the NSCBC whereas all the variables in the nonreflecting BC are "soft" and are changing in the simulation according to the characteristic waves across the boundary.

The NSCBC is based on an analysis of the characteristic waves of the Euler equations. Using the characteristic analysis, the Euler equations can be recast into a new form at the flow boundary by recalculating the hyperbolic terms using the characteristic wave amplitudes (Thompson 1987; Poinsot and Lele 1992). Although the concept of "characteristic lines" may be questionable for the Navier–Stokes equations, it is logical to assume that waves for the Navier–Stokes equations are associated only with the hyperbolic part of the equations. In the NSCBC, flow variables at the boundary are simply obtained as the solution of the modified equations by replacing the hyperbolic terms using the characteristic wave amplitudes, which serve as the "numerical" or "soft" boundary

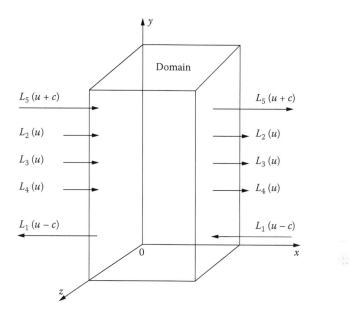

$L_5 (u + c)$

$L_2 (u)$

$L_3 (u)$

$L_4 (u)$

$L_1 (u - c)$

$L_5 (u + c)$

$L_2 (u)$

$L_3 (u)$

$L_4 (u)$

$L_1 (u - c)$

FIGURE 2.1 Waves entering and leaving the computational domain in the direction for a subsonic flow.

condition for the simulation. Considering only the x direction for simplicity, the waves entering and leaving the computational domain for a subsonic flow are shown in Figure 2.1. The wave amplitudes (Thompson 1987; Poinsot and Lele 1992) for this subsonic flow can be specified as

x direction:
$$
\begin{cases}
\mathcal{L}_1 = (u-c)\left(\dfrac{\partial p}{\partial x} - \rho c \dfrac{\partial u}{\partial x}\right) \\[2mm]
\mathcal{L}_2 = u\left(\dfrac{\partial p}{\partial x} - c^2 \dfrac{\partial \rho}{\partial x}\right) \\[2mm]
\mathcal{L}_3 = u\dfrac{\partial v}{\partial x} \\[2mm]
\mathcal{L}_4 = u\dfrac{\partial w}{\partial x} \\[2mm]
\mathcal{L}_5 = (u+c)\left(\dfrac{\partial p}{\partial x} + \rho c \dfrac{\partial u}{\partial x}\right)
\end{cases}
\tag{2.2}
$$

Similarly, the wave amplitudes for subsonic flows in the y direction and z direction are given as

$$y \text{ direction}: \begin{cases} \mathcal{L}_1 = (v - c)\left(\dfrac{\partial p}{\partial y} - \rho c \dfrac{\partial v}{\partial y}\right) \\[2mm] \mathcal{L}_2 = v\dfrac{\partial u}{\partial y} \\[2mm] \mathcal{L}_3 = v\left(\dfrac{\partial p}{\partial y} - c^2 \dfrac{\partial \rho}{\partial y}\right) \\[2mm] \mathcal{L}_4 = v\dfrac{\partial w}{\partial y} \\[2mm] \mathcal{L}_5 = (v + c)\left(\dfrac{\partial p}{\partial y} + \rho c \dfrac{\partial v}{\partial y}\right) \end{cases} \tag{2.3}$$

and

$$z \text{ direction}: \begin{cases} \mathcal{L}_1 = (w - c)\left(\dfrac{\partial p}{\partial z} - \rho c \dfrac{\partial w}{\partial z}\right) \\[2mm] \mathcal{L}_2 = w\dfrac{\partial u}{\partial z} \\[2mm] \mathcal{L}_3 = w\dfrac{\partial v}{\partial z} \\[2mm] \mathcal{L}_4 = w\left(\dfrac{\partial p}{\partial z} - c^2 \dfrac{\partial \rho}{\partial z}\right) \\[2mm] \mathcal{L}_5 = (w + c)\left(\dfrac{\partial p}{\partial z} + \rho c \dfrac{\partial w}{\partial z}\right) \end{cases} \tag{2.4}$$

In Equations (2.2)–(2.4), c represents the sonic speed and other variables have their usual meanings.

The characteristic waves play a significant role in NSCBC and nonreflecting BC, and they can be classified as incoming waves entering the computational domain and outgoing waves leaving the computational domain as shown in Figure 2.1. From a numerical point of view, outgoing waves leaving the computational domain do not cause any problem because their amplitudes can be readily calculated from the solution inside the computational domain. In the meantime, incoming waves entering the computational domain can be problematic because they rely on the solution outside the computational domain, which is an unknown in the simulation.

In the nonreflecting characteristic boundary condition (e.g., Thompson 1987), the amplitudes of all the incoming waves are simply set to zero. As a consequence, the unknown field outside the computational domain is excluded from the simulation. Computational practice has shown that nonreflecting BC is very effective in controlling the wave reflections from outside the computational domain. Therefore, nonreflecting BC have been successfully utilized in DNS and LES of many compressible flows. Applying the nonreflecting BC normally leads to a smooth flow field near the boundary. However, setting the amplitudes of all the incoming waves to zero is not physically very sound. When a nonreflecting BC is used, not all the flow variables at the boundary are fixed and they are changing with time according to the characteristic waves across the boundary, which may not be preferable in some cases such as for an inflow where the flow variables are physically fixed. For instance, applying the nonreflecting inflow BC to a flow with buoyancy effect can be disastrous, as the drifting density at the inflow can lead to an unphysical buoyancy effect at the inflow.

The NSCBC can overcome the drawback of the nonreflecting BC by incorporating the physical conditions into the formulation of the characteristic BC, using the concept of the local one-dimensional inviscid (LODI) relations (Poinsot and Lele 1992). The LODI relations provide compatible relations between the physical boundary conditions and the amplitudes of the characteristic waves crossing the boundary. Assuming the flow is locally one-dimensional in the direction normal to the boundary, the $\mathcal{L}_i, i = 1, \ldots, 5$, defined in Equations (2.2)–(2.4), of the incoming waves can be calculated from the amplitudes of the outgoing waves in the LODI relations; this provides a much better approximation of the characteristic waves across the boundary than the nonreflecting BC. The LODI relations are easy to implement and can be viewed as compatibility relations between the choices made for the physical boundary conditions and the amplitudes of waves crossing the boundary. It is worth noting that the values obtained for the wave amplitude variations through the LODI relations are approximate since the complete Navier–Stokes equations involve viscous terms and terms for other directions. Nevertheless, the LODI relations are the most important features of the NSCBC. As an example, the

LODI relations for density, pressure, and velocity components of a subsonic flow are given as

x direction:
$$
\begin{cases}
\dfrac{\partial \rho}{\partial t} + \dfrac{1}{c^2}\left[-\mathcal{L}_2 + \dfrac{1}{2}(\mathcal{L}_5 + \mathcal{L}_1) \right] = 0 \\[2ex]
\dfrac{\partial p}{\partial t} + \dfrac{1}{2}(\mathcal{L}_5 + \mathcal{L}_1) = 0 \\[2ex]
\dfrac{\partial u}{\partial t} + \dfrac{1}{2\rho c}(\mathcal{L}_5 - \mathcal{L}_1) = 0 \\[2ex]
\dfrac{\partial v}{\partial t} + \mathcal{L}_3 = 0 \\[2ex]
\dfrac{\partial w}{\partial t} + \mathcal{L}_4 = 0
\end{cases}
\tag{2.5}
$$

y direction:
$$
\begin{cases}
\dfrac{\partial \rho}{\partial t} + \dfrac{1}{c^2}\left[-\mathcal{L}_3 + \dfrac{1}{2}(\mathcal{L}_5 + \mathcal{L}_1) \right] = 0 \\[2ex]
\dfrac{\partial p}{\partial t} + \dfrac{1}{2}(\mathcal{L}_5 + \mathcal{L}_1) = 0 \\[2ex]
\dfrac{\partial u}{\partial t} + \mathcal{L}_2 = 0 \\[2ex]
\dfrac{\partial v}{\partial t} + \dfrac{1}{2\rho c}(\mathcal{L}_5 - \mathcal{L}_1) = 0 \\[2ex]
\dfrac{\partial w}{\partial t} + \mathcal{L}_4 = 0
\end{cases}
\tag{2.6}
$$

z direction:
$$
\begin{cases}
\dfrac{\partial \rho}{\partial t} + \dfrac{1}{c^2}\left[-\mathcal{L}_4 + \dfrac{1}{2}(\mathcal{L}_5 + \mathcal{L}_1) \right] = 0 \\[2ex]
\dfrac{\partial p}{\partial t} + \dfrac{1}{2}(\mathcal{L}_5 + \mathcal{L}_1) = 0 \\[2ex]
\dfrac{\partial u}{\partial t} + \mathcal{L}_2 = 0 \\[2ex]
\dfrac{\partial v}{\partial t} + \mathcal{L}_3 = 0 \\[2ex]
\dfrac{\partial w}{\partial t} + \dfrac{1}{2\rho c}(\mathcal{L}_5 - \mathcal{L}_1) = 0
\end{cases}
\tag{2.7}
$$

The applications of the LODI relations in the NSCBC are straightforward. Considering the inflow in the x direction shown in Figure 2.1, for instance, the incoming waves are $\mathcal{L}_5$, $\mathcal{L}_2$, $\mathcal{L}_3$, $\mathcal{L}_4$, while the only outgoing wave is $\mathcal{L}_1$. Assuming that the "soft" variable is density, the pressure varies according to $p = \rho R T$, and the velocity components are fixed constants at the inflow, $\mathcal{L}_5$, $\mathcal{L}_2$, $\mathcal{L}_3$, and $\mathcal{L}_4$ can then be calculated using the LODI relations given in Equation (2.5) as $\mathcal{L}_5 = \mathcal{L}_1$, $\mathcal{L}_2 = \mathcal{L}_1$, $\mathcal{L}_3 = \mathcal{L}_0$, and $\mathcal{L}_4 = 0$. Using these wave amplitudes, the governing equations at the boundary can be recast into physically meaningful expressions, which will provide numerical boundaries for the simulation. Depending on the physical conditions at the boundary, there can be a variety of LODI relations, as discussed by Poinsot and Lele (1992). The LODI relations can be viewed as compatibility relations between the choices made for the physical boundary conditions and the amplitudes of waves crossing the boundary. It needs to be mentioned that values obtained for the wave amplitude variations through LODI relations are approximate because the complete Navier–Stokes equations involve viscous and parallel terms.

The NSCBC method has proved to be very effective in the specification of boundary conditions for DNS and LES of compressible flows. It is often used in association with the high-order nondissipative numerical algorithms (Lele 1992). The NSCBC uses the correct number of boundary conditions required for well-posedness of Navier–Stokes equations that can avoid numerical instabilities and spurious wave reflections at the computational boundaries. This is very important to DNS and LES, where high-order nondissipative numerical schemes are often employed to ensure numerical accuracy. In DNS or LES using high-order numerical schemes, there can be severe consequences if the well-posedness of Navier–Stokes equations is not guaranteed. For instance, numerical oscillations and spurious wave reflections will occur near the inflow boundary if all the variables are fixed. These may not cause problems in a numerical simulation using lower-order dissipative numerical schemes such as those in RANS because the dissipative error in the scheme may smooth the flow and avoid numerical instabilities, but at the expense of low accuracy. The NSCBC method allows a nonreflecting treatment for waves approaching the boundary at normal incidence. For multidimensional problems where the waves do not reach the boundary at normal incidence, the NSCBC treatment leads to small levels of reflection but still prevents oscillations and ensures well-posedness. To ensure better performance of the NSCBC, the flow boundaries may be placed in locations where the flow is more or less one dimensional, or a sponge layer can be used to make the flow be one dimensional such as that associated with the outflow boundary.

In summary, two types of inflow/outflow boundary conditions have to be provided to solve numerically the compressible flow governing equations: physical conditions and "soft" numerical conditions required by the numerical solution of the governing equations. In the NSCBC method, the physical and "soft" numerical conditions are interlinked through the LODI relations. The principles and assumptions of the NSCBC are summarized as follows:

- Physical conditions are specified according to well-posedness studies of the Navier–Stokes equations. Variables that are not imposed by physical boundary conditions are computed on the boundaries by solving modified governing equations using the characteristic wave crossing the boundary.

- The waves for the Navier–Stokes equations are associated only with the hyperbolic part of the Navier–Stokes. (Although the Navier–Stokes equations are not hyperbolic, they can be assumed to propagate waves like the Euler equations.) All incoming wave amplitudes at a given boundary can be estimated from the original choice of the physical boundary conditions imposed on this boundary and can be expressed in terms of the outgoing wave amplitudes.

The NSCBC treatment for compressible flows discussed above is not restricted to inflow and outflow boundaries. As a general formation strategy of boundary conditions, it can be used in the specification of wall boundaries and open boundaries as well. So far the discussion on boundary conditions has been restricted to nonreacting flows. Characteristic boundary conditions can also be developed for reacting flows involving multiple species and chemical reactions (e.g., Baum et al. 1995; Sutherland and Kennedy 2003). Details of these boundary conditions can be found in the literature.

D. Inflow/Outflow BC for Incompressible Flows

The characteristic boundary conditions for compressible flows discussed above are not relevant to incompressible flows. In CFD solvers for compressible flows, density is one of the variables in the solution while pressure is directly calculated from the density and temperature. However, CFD solvers of incompressible flows are normally pressure based, in which pressure is solved from a governing equation such as Poisson's equation for pressure. The pressure is a somewhat peculiar quantity in incompressible flows. It is not a thermodynamic variable as there is no "equation of state" for an incompressible fluid. Its gradient is important in determining the velocity field. In an incompressible flow, the pressure propagates at infinite speed (or, in other words, the sonic speed is infinite) in order to keep the flow incompressible; that is, the pressure is always in equilibrium with a time-varying, divergence-free velocity field. It is also often difficult and/or expensive to compute. The general discussions on inflow and outflow BC for DNS and LES presented before are relevant to incompressible flows, apart from the NSCBC for compressible flows. For incompressible flows, the central issue of boundary conditions is associated with the boundary conditions for Poisson's equation for pressure, or pressure Poisson equation (PPE).

For incompressible flows, the fundamental governing equations lack an independent equation for the pressure. The pressure is governed by the PPE, which can be derived from the continuity and momentum equations, as given in Equation (1.21). It can be shown that the PPE is an elliptic problem; that is, pressure values or its gradients on boundaries must be known in order to compute the whole flow field. In applying boundary conditions to the PPE, it is important to ensure that zero divergence of the velocity as shown in Equation (1.16) is enforced. There have been considerable discussions in the literature regarding the proper boundary condition for the pressure equation (e.g., Nordström et al. 2007; Sani et al. 2006). The boundary conditions by Nordström et al. (2007) have the same form on both inflow and outflow boundaries and lead to a divergence-free solution. Both Neumann and Dirichlet boundary conditions can be applied to the PPE. A Neumann condition can be derived simply by applying the normal component of the momentum equation at the boundary (Gresho and Sani 1987), while a Dirichlet boundary condition can be obtained by integrating the tangential component of the momentum equation along the boundary, which should give the same solution (Abdallah and Dreyer 1998). The boundary conditions and solvers for the PPE are an important part of the numerical solution of an incompressible flow.

II. WALL BOUNDARY CONDITIONS

Wall boundary conditions play a dominant role in high-quality DNS and LES of wall-bounded flows. The derivation of appropriate wall boundary conditions is not a trivial task, and has implications in computational costs of the simulation and accuracy of the numerical results. Wall boundaries are encountered in almost all of the practical applications of fluid flow, heat transfer, and combustion applications. The presence of walls influences the flow dynamics in a significant manner. In the near-wall region, there is normally a sharp gradient for flow variables such as velocities, leading to larger shear and normal stresses compared to the flow regions far away from the wall boundary. Due to the large viscous stresses in the near-wall region, turbulence generation and transportation are also significantly affected. The near-wall flow and its control are very important to a broad range of practical applications. For instance, control of the level of heat fluxes to walls is of crucial importance to the lifetime of the practical device such as the combustor of a gas-turbine system and the cylinder of a combustion engine. Although near-wall flow in very important, it has not been fully understood, especially for turbulent flows. The near-wall flow involves a rapid change of the turbulent time and length scales. In the near-wall flow region, the generation and transportation of turbulence are more significant compared with regions away from the wall. In CFD applications, wall boundary conditions have to be properly set up so that the flow in the near-wall region can be represented as faithfully as possible.

In practice, the wall boundary conditions for DNS and LES are not significantly different from those for RANS. Although DNS and LES need to be able to represent the unsteadiness of turbulence, which is different from the traditional CFD based on RANS approach, the unsteadiness is mainly embedded in the time-dependent governing equations and the inflow boundary conditions. Under a time-dependent solver with an unsteady upstream condition, temporally and spatially resolved flow fields can be obtained for the flow in the near-wall region in DNS and LES. For instance, the maximum heat fluxes through the wall that control the maximum local thermal load imposed on materials can be obtained from DNS and LES. However, traditional RANS CFD cannot predict the instantaneous maximum heat flux, but only the mean values.

Wall boundary conditions are an integral part of the numerical simulation, of which the accuracy is closely related to the grid resolution in the near-wall region as well as the resolutions in the flow regions away from the wall. There is a difference between the wall boundary treatment for DNS and LES. In a DNS simulation, the near-wall flow has to be resolved fully, which often requires a very fine mesh in the near-wall region. In an LES approach, this may not be necessary and a modeling strategy similar to that used in RANS is often employed to avoid fully resolving the near-wall flow. Pope (2000) classified LES of wall-bounded flows as (1) large-eddy simulation with near-wall resolution (LES-NWR), where the filter and grid are sufficiently fine to resolve 80% of the energy everywhere; and (2) large-eddy simulation with near-wall modeling (LES-NWM), where the filter and grid are sufficiently fine to resolve 80% of the energy remote from the wall, but not in the near-wall region. In the meantime, the simulation can be regarded as very-large-eddy simulation (VLES) if the filter and grid are too coarse to resolve 80% of the energy.

In the near-wall region, flow properties can change rapidly and the most notable one is velocity. Due to the friction at the wall, velocity can be zero on the wall surface. Associated with this nonslip condition, large velocity gradients usually exist in the direction normal to the wall boundary and are mainly responsible for the production of turbulent kinetic energy in the near-wall region. In DNS or LES, the numerical prediction has to give an accurate representation of these gradients, which is often difficult due to the excessive requirement on mesh density. For a turbulent flow, an extremely fine grid is required not only in the wall-normal direction, but also in all spatial directions in order to resolve the near-wall turbulence, including coherent structures such as the well-known high- and low-speed streaks (Breuer 2007; Piomelli and Balaras 2002). For model-free DNS, there is no option but to fully resolve the flows in the near-wall region. The wall boundary condition formulation in DNS can simply follow the methods described earlier in this chapter. For example, the wall boundary conditions for DNS of compressible flows can be specified by using the NSCBC method, taking into account the possible physical conditions in the wall boundary such as nonslip conditions and adiabatic or constant-temperature walls.

In LES, the near-wall flow can either be resolved as in DNS or modeled as in a traditional RANS modeling approach. In terms of computational costs, it is normally prohibitive to perform LES-NWR to resolve the near-wall flows for many practical problems. Near-wall models therefore play an important role in LES of practical problems. However, developing suitable near-wall models for LES is a difficult task due to the complexity of near-wall turbulent flows. LES-NWM is still a developing area. Intensive research activities are ongoing to widen the applicability of LES, partially leading to hybrid LES-RANS approaches. This will result in exciting new techniques and prospects for LES. Although there have been a significant amount of and continuous efforts in developing appropriate near-wall models for LES, the current near-wall models used in LES are predominately taken from those approaches used in the traditional RANS approach.

For simulations of near-wall flows using RANS and LES with near-wall modeling, an important parameter is the wall unit, which is defined as

$$y^+ = \frac{u_\tau y}{\nu} \tag{2.8}$$

where the friction velocity (a characteristic velocity at a wall defined from the dimensional analysis) is defined as $u_\tau = \sqrt{\tau_w/\rho}$ with τ_w standing for the wall shear stress and ρ for the fluid density at the wall, y for the distance to the nearest wall, and ν for the local kinematic velocity of the fluid. The dimensionless wall unit y^+ (often referred to simply as y plus) represents the dimensionless wall distance and is commonly used in boundary layer theory for walls and in defining the law of the wall.

In DNS and LES, both the accuracy of the numerical scheme and the mesh size are of importance to the accuracy of the results. In general, a coarser mesh may be used if a higher-order numerical scheme is employed. For a given numerical scheme, the mesh size used in a practical simulation is the major factor determining the accuracy of the results. Based on a private communication with Sagaut in 2004, Breuer (2007) discussed the typical mesh sizes for numerical simulations of near-wall turbulent flows using DNS, wall-resolved LES, and LES with an appropriate wall model. Table 2.1 shows these typical mesh sizes expressed in wall units. In DNS and LES of near-wall flows, the quality of the results can exhibit a dependence on the size of mesh, especially in LES when the resolution in the near-wall region is not very high. Furthermore, the resolutions in all three directions are important. As discussed by Breuer (2007), the resolutions in

TABLE 2.1 Typical Mesh Size in Wall Units for a Turbulent Boundary Layer Flow Using DNS, Wall-Resolved LES, and LES with an Appropriate Wall Model

	DNS	Wall-Resolved LES	LES with Wall Model
Streamwise Δx^+	10–15	50–150	100–600
Spanwise Δz^+	5	10–40	100–300
Wall-normal (Δy^+)	1	1	30–150
Number of points in $0 < y^+ < 10$	3–5	3–5	—

Source: Breuer (2007) based on the private communication with Sagaut in 2004. With permission from Cambridge University Press.

the streamwise direction Δx^+ and in the spanwise direction Δz^+ are also very important parameters that govern the quality of the LES solution:

- Poor-resolution LES with large Δx^+ and Δz^+ (measured in wall units), that is, $\Delta x^+ \geq 100$ and $\Delta z^+ \geq 30$, leading to unphysical streaks and large error on the skin friction

- Medium-resolution LES with moderate Δx^+ and Δz^+, that is, $50 \leq \Delta x^+ < 100$ and $12 \leq \Delta z^+ < 30$, leading to thicker and shorter streaks and error on the skin friction

- High-resolution LES with small Δx^+ and Δz^+, that is, $\Delta x^+ < 50$ and $\Delta z^+ < 12$, leading to good agreement of the predicted skin friction in plane channel compared with DNS or experiments when nondissipative numerical methods are used

For practical flows with high Reynolds numbers, an extremely fine resolution is always required, which is very often not achievable. In order to overcome this problem, near-wall models or wall functions bridging the near-wall region, which enables the first grid point to be placed in the region of $30 < y^+ < 50$ within the logarithmic part of the velocity profile (see the next section, "Classical Wall Models"), have been developed. Consequently, the near-wall flow behavior is not resolved in detail, which also leads to reduced requirements with respect to the grid characteristics in streamwise and spanwise directions. As a result, the resolution requirements are drastically reduced, allowing high Reynolds number flows to be tackled. Since the classical wall models were the basis of near-wall flow modeling, they are briefly reviewed as follows.

A. Classical Wall Models

The classical wall models were developed based on the analysis of boundary layer flows. In the motion of a fluid above a wall surface, the influence of viscosity is mainly confined to a boundary layer close to the wall surface. A diagram showing a typical wall boundary layer flow is shown in Figure 2.2. Within the wall boundary layer, the velocity changes rapidly from zero at the wall (no-slip condition) to the free stream velocity. Practically, the boundary layer thickness δ may be defined as the distance from the solid surface where the local velocity u is, say, 99% of the free stream velocity U. Experimental studies of a turbulent wall boundary layer suggest that it may be divided into two regions (or layers): the inner

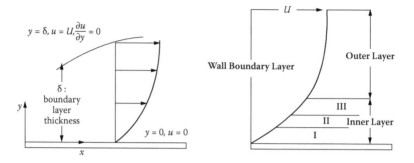

FIGURE 2.2 Schematic of a wall boundary layer flow.

(wall) layer approximately at $0 < y < 0.2\delta$ and an outer layer at $y > 0.2\delta$. The velocity distribution in the wall layer can be analyzed by using dimensional analysis, which led to the law of the wall. One of the first wall models developed and applied to flows in plane channels and annuli was by Schumann (1975). It is based on the phase coincidence of the instantaneous wall shear stress $\tau_{xy,w}$ and the tangential velocity component u at the first grid point nearest the wall. The coincidence assumed between the wall shear stress $\tau_{xy,w}$ and the velocity component u has been experimentally verified. In the classical wall models, a dimensionless velocity u^+ can be defined as

$$u^+ = \frac{u}{u_\tau} \tag{2.9}$$

The law of the wall states that u^+ is a function of y^+ only. At the first grid point nearest to the wall, it can be shown that $u^+ = y^+$. As determined by the wall distance, a corresponding law for the viscous sublayer (VS: region I of the inner layer in Figure 2.2), the logarithmic buffer layer (BL: region II of the inner layer in Figure 2.2), and the logarithmic outer layer (OL: region III of the inner layer in Figure 2.2) is assumed:

$$u^+ = y^+ \qquad \text{for the VS } 0 \le y^+ < 5 \tag{2.10}$$

$$u^+ = a_2 \ln(y^+) + b_2 \qquad \text{for the BL } 5 \le y^+ < 30 \tag{2.11}$$

$$u^+ = a_3 \ln(y^+) + b_3 \qquad \text{for the OL } 30 \le y^+ < 500 \tag{2.12}$$

where a_2, a_3, b_2, and b_3 are empirical constants, which may be given as $a_2 = 5.0$, $a_3 = 2.5$, $b_2 = -3.05$, and $b_3 = 5.0$–5.2. Relations such as Equations (2.10)–(2.12) can be conveniently implemented into a CFD code, avoiding a numerical solution of the flow in the wall boundary layer, which normally incurs very heavy computational load due to the existence of large gradients in the near-wall flow region. The classical law of the wall holds for smooth surfaces under zero and moderate pressure gradients. The presence of an adverse pressure gradient is normally responsible for deviations from this law.

There are several enhanced versions of the classical law of the wall. Based on Schumann's approach (1975), Piomelli et al. (1989) took into account the inclination of the near-wall structures and the resulting temporal delay between the tangential velocity and the wall shear stress. Another wall model suggested by Piomelli et al. (1989) was the ejection boundary condition, which was based on the observation that the near-wall dynamics were dominated by sweeps and ejections. Both models led to slightly improved results for the plane channel flow compared with Schumann's original formulation.

The classical law of the wall models, including the enhanced versions, have been successfully used in many simple wall boundary layer flows. A major drawback of all these models is that they are difficult to assign to complex, statistically three-dimensional flows because they require the determination of the averaged wall shear stress and velocity. In addition, the use of the customarily applied laws of the wall is highly questionable for flows in complex geometries involving large pressure gradients or local separation and recirculation regions. In complex geometries such as a flow configuration involving "corners," the boundary layer assumption is no longer valid.

The analysis of the wall boundary layer flows leading to the law of the wall was based on the averaged flows, that is, time- or ensemble-averaged flow quantities. An analytical expression for the local wall shear stress was derived by Hoffmann and Benocci (1995), where the boundary layer equations were analytically integrated, coupled with an algebraic eddy-viscosity model. Neglecting the convection terms and approximating the unsteady term, an expression for the wall shear stress was obtained, leading to satisfactory results for plane channel flows and rotating channel flows. An improved model was also derived by Manhart (2001), who took the local instantaneous pressure gradient in the streamwise direction into account.

There were also wall models based directly on the near-wall velocity fields. Extended from Schumann's and Piomelli's models, Bagwell et al.

(1993) used the entire velocity field in a plane parallel to the wall in order to determine the wall shear stress, on the basis of a linear stochastic estimation approach. Following Schumann's concept, Werner and Wengle (1993) suggested a wall model that is also based on the phase coincidence but that applies the laws of the wall directly to the instantaneous velocity field. This simplifies the determination of the wall shear stress and allows the use of their model in flows with separations. However, the application of the laws of the wall to instantaneous velocities and in separated flows is theoretically not justified. The application of these models becomes questionable.

The limitations of the classical law of the wall models for the near-wall flow are associated with the assumptions made in developing these models, most notably the boundary layer flow assumption. There have been many successful applications of these wall models to RANS-based CFD. However, strictly speaking, these wall models are only applicable to simple flow configurations, and usually they are only applicable to the mean flow quantities. Attention is needed when these laws are used together with LES, which in principle should be time dependent and fully three-dimensional. A mismatch would occur if these wall models were used directly in LES. To overcome this problem, a more suitable approach is the zonal or nonzonal approach, which try to combine RANS-based near-wall modeling together with LES for the flow region away from the wall boundaries, as discussed in the section that follows.

B. Zonal and Nonzonal Approaches for LES

Since there is a lack of well-established near-wall models for LES, a natural option is to use those models developed for RANS. A wall-bounded flow can be divided into two regions: a near-wall region where the effect of wall boundary on the fluid dynamics is significant, which is also referred to as the inner layer; and a region away from the wall where the presence of the wall boundary does not have a significant effect on the flow, which is also referred to as the outer layer. A possibility for LES of fluid flows with wall modeling is to use so-called zonal or nonzonal approaches based on the explicit solution of a different set of equations in the inner and outer layers (Balaras and Benocci 1994; Balaras et al. 1996; Cabot and Moin 1999; Piomelli and Balaras 2002). This modeling approach takes advantage of using RANS for the near-wall region; therefore, it may avoid the difficulties of LES for the near-wall region. The basic assumption is that the interaction between the near-wall region and the outer region is weak.

In a zonal or nonzonal approach, simplified governing equations such as the two- and three-dimensional thin boundary layer equations or the RANS equations with a statistical turbulence model can be used for modeling the inner near-wall region. These equations can be solved on an embedded inner grid in the direct vicinity of the wall, whereas the original LES prediction can be carried out on an outer grid not including the near-wall region. One example of this approach is a two-layer model proposed by Balaras and Benocci (1994) and extensively tested by Balaras et al. (1996). In this two-layer model, the boundary layer equations for the inner layer can be given as

$$\frac{\partial \overline{u}_i}{\partial t} + \frac{\partial}{\partial x_i}(\overline{u}_n \overline{u}_i) = -\frac{\partial \overline{p}}{\partial x_i} + \frac{\partial}{\partial x_n}\left[(v+v_t)\frac{\partial \overline{u}_i}{\partial x_n}\right] \qquad (2.13)$$

where n denotes the wall-normal direction y and $i = 1,2$ or $i = 1,3$ depending on whether the wall plane is in the x–y plane or x–z plane. The wall-normal velocity component u_n is computed based on the mass conservation for the inner layer, which has several grid points, each with its own value of u_n. The system is closed by setting the wall boundary condition such as the no-slip condition for the wall-side boundary of the inner layer, and by setting the velocity obtained from the outer-flow LES prediction as a "freestream" condition at the outer-side boundary of the inner layer (Piomelli and Balaras 2002). Furthermore, the pressure gradient $\partial \overline{p} / \partial x_i$ in Equation (2.13) is assumed to be independent of y in the inner layer and thus taken from the outer-flow prediction. Consequently, no Poisson equation for the pressure of an incompressible flow has to be solved and the costs for solving the two momentum equations in the inner layer are only marginally higher than using equilibrium boundary conditions such as in the wall models of Schumann (1975) and Piomelli et al. (1989). In this approach, the quality of the simulation depends strongly on the choice of the model for the eddy viscosity v_t. The most commonly used model for v_t, originally applied by Balaras and Benocci (1994) and Balaras et al. (1996), is a simple mixing-length model with near-wall damping. It can be given as

$$v_t = (\kappa y)^2 D(y)|\overline{S}| \qquad (2.14)$$

where κ denotes the von Kármán constant, $|\overline{S}|$ is the magnitude of the resolved strain rate, $D(y) = 1 - \exp[-(y^+/A^+)^3]$ is a damping function that ensures the correct behavior of v_t at the wall with y^+ representing the distance from the wall in wall units and $A^+ = 25$ as a constant. Finally, the wall stress components obtained from integrating Equation (2.13) in the inner layer are used as boundary conditions for the outer LES prediction.

This two-layer model was successfully applied to channel flows with and without extra rotation and backward-facing step flows. In the backward-facing case, however, the boundary layer equations used within that model are no longer valid in the vicinity of the separation region. Nevertheless, the full time- or ensemble-averaged Navier–Stokes equations can always be used in the inner layer when the boundary layer assumption is not valid. If the full RANS equations are applied instead, a hybrid LES–RANS approach is achieved. Since the regions for RANS and LES are defined in advance in the numerical simulation, this method is referred to as a zonal approach.

In companion to the zonal approach, the counterpart to the zonal technique discussed above is the nonzonal hybrid LES–RANS approach. In a nonzonal approach, a gradual transition between both LES and RANS takes place based on an automated switch, ideally removing the need for user-defined information. Speziale's (1998) formulation belongs to this hybrid concept. Following the idea of Speziale (1998), Zhang et al. (2000) numerically demonstrated this hybrid concept for a flat-plate boundary layer with and without separation, which was perhaps the first successful application of the concept in practical simulations. Conceptually, their so-called flow simulation methodology is very similar to the detached-eddy simulation (DES) proposed by Spalart et al. (1997) and Spalart (2000), which is more widely known in the field. The DES approach may be considered a zonal method because the LES and RANS domains are fully determined by the grid topology and the segmentation is independent of the flow solution.

The DES has been a great success in practical simulations of wall-bounded flows. In DES, the attached flow regions near the walls are distinguished from the separated flow regions with detached eddies. The near-wall flow is properly predicted based on RANS with statistical turbulence models, whereas the detached flow region, including the large-scale unsteady vortical structures, are computed more reasonably by LES. The basic concept is to combine the advantages of both methods, yielding an optimal solution at least for a special class of flows, and to afford predictions of high Reynolds number flows with reasonable computational efforts. The DES method could be regarded as a natural hybrid method combining RANS and LES. It means that, near solid boundaries, the governing equations work in the RANS mode where all turbulent stresses are modeled using the traditional RANS turbulence models, while far away from solid boundaries, the method switches to the LES mode. Note that pressure and velocity fields are time- or ensemble-averaged in the near-wall region. Therefore, the unsteady vortical structures in the near-wall region are not resolved directly and DES is not able to give detailed information on the near-wall dynamic structures.

Although DES seems to be a useful technique for predictions of high Reynolds number flows, a variety of open issues need to be addressed before one can rely on such a hybrid method. These include, in particular, the demand for appropriate coupling techniques between LES and RANS, adaptive control mechanisms, and proper SGS-RANS turbulence models (Breuer 2007). In a hybrid approach, the quality of the numerical results depends on both the LES and RANS and their coupling. The final numerical results rely partially on the wall boundary conditions implemented in the approximate RANS modeling in the near-wall region. The classical wall functions discussed above are examples of RANS modeling in the near-wall region, which can be used in the hybrid approaches. Piomelli and Balaras (2002) gave a more complete review of wall-layer models for LES, which can be referred to for more information.

The near-wall models discussed so far have been restricted to velocities. Appropriate boundary conditions for the pressure field (if required at all) are not so critical. They depend on the flow problem considered and on the numerical methodology applied. For example, the boundary layer over a flat plate at rest exhibits a zero wall-normal pressure gradient at the wall, which can be discretized by a Neumann boundary condition. For a more general case of a flow over a curved, moving, or rotating surface or when external forces such as buoyancy or centrifugal forces are present, large pressure gradients may appear in the vicinity of the surface. In that case it is advisable to determine the pressure gradient based on a simplified momentum equation in the wall-normal direction using one-sided finite differences or to extrapolate the pressure at the wall from the internal region to the boundary. Because these techniques used in the context of LES do not deviate from those used in the traditional RANS approach, one may refer to the basic literature for CFD.

C. Near-Wall Models for Nonisothermal Flows

The wall boundary conditions and near-wall models discussed so far are mostly sufficient for isothermal flows. For such flows, there is normally no need to solve for an energy equation and therefore there is no need to deal with boundary conditions for temperature, or internal energy or enthalpy. In many heat transfer and reacting flow applications, an energy equation for temperature, or internal energy or enthalpy, has to be solved and one has to deal with the boundary condition of the energy equation. In this case, a situation similar to that of the velocity field arises. In principle, a wall heat flux or wall temperature can be prescribed and discretized without further approximations. Indeed, this has to be the case for DNS and LES–NWR (large-eddy simulation with near-wall resolution). Like the viscous wall layer for the velocity field, this measure requires the conductive wall layer to be resolved. As determined by the molecular Prandtl number of the fluid, which describes the ratio of diffusivities for momentum and heat, this layer can be even thinner than the viscous layer for many fluid materials such as liquids with Prandtl number greater than one. This implies that even finer mesh will be needed to resolve the near-wall layer of temperature than that of the velocity.

In some LES applications of nonisothermal flows, a very fine near-wall resolution is not possible or not desired. In such LES–NWM (large-eddy simulation with near-wall modeling), near-wall models for heat transfer are needed. These models are basically analogous to the wall models for the velocity field such as the models of Schumann (1975) and Piomelli et al. (1989) described in previous sections. Grötzbach (1981; 1987) and Grötzbach and Wörner (1999) proposed a time-dependent formulation of wall models for the temperature equation. These models relate the instantaneous local heat fluxes to the temperature fluctuations at the grid point nearest to the wall by using time-averaged wall laws. In the context of flame/wall interaction, Poinsot and Veynante (2001) discussed the law of the walls for nonisothermal flows. Jiang et al. (2007b) attempted to develop a law of the walls for nonisothermal flows using DNS results of an impinging jet flame. Nevertheless, the law of the walls for nonisothermal flows are much less developed than those for isothermal flows. There is a need to further develop such models. Clearly, the near-wall models for chemically reacting flows such as combustion problems are more complex than those for nonreacting isothermal flows. In a reacting flow, the transient effects are more significant in the near-wall region because the flame may extinguish in the vicinity of the walls. In addition, wall boundary conditions are needed for the concentrations of chemical species. The additional scalar transport equations for chemical species can be treated in much the same way as the energy equation for temperature. However, the diffusivities for different chemical species can be significantly different, and also different from those for the momentum and heat. Consequently, both the near-wall resolution and the near-wall modeling of chemically reacting flows are very challenging, and normally require much finer mesh and/or greater modeling care than those for nonreacting flows.

III. OTHER BOUNDARY CONDITIONS

Inflow and outflow boundary conditions and wall boundary conditions play a dominant role in DNS and LES predictions of high quality. Intensive research activities have been carried out to develop inflow/outflow and wall boundary conditions. However, other types of boundary conditions are also often encountered in a practical DNS or LES calculation, or generally speaking, in any CFD problem. One example is the artificial boundaries arising owing to finite computational domains for open boundary flow problems. Other examples include periodic and symmetry boundary conditions. The derivation of these boundary conditions is not a trivial task, because it also needs to represent or approximate the physical conditions at the boundary locations as faithfully as possible. In the following sections, appropriate far-field boundary conditions associated with the artificial boundaries arising owing to finite integration domains for the prediction of external compressible flows is presented first. This issue is directly related to the boundary conditions applied to acoustic simulations, where DNS and LES are often very useful. Periodic and symmetry boundary conditions are then discussed in the context of high-order numerical schemes for DNS and LES. Finally, other relevant issues in the specification of boundary conditions are briefly discussed.

A. Far-Field and Open Boundary Conditions for Compressible Flows

DNS and LES are particularly useful in computational acoustics for compressible flows, due to the fact that they are time-accurate (in comparison with the traditional RANS modeling approach) so that they can predict the time-dependent pressure fluctuations constituting the acoustic field. Far-field boundary conditions for compressible flows are an essential part of the boundary conditions for numerical predictions of the acoustic field, which normally require a relatively large computational domain. These far-field boundary conditions are artificial boundaries, locating in regions where the flow does not experience significant changes. However, they are still very important to the overall quality of the numerical solution. Inappropriate far-field boundary conditions can severely pollute the solutions inside the computational domain, leading to unphysical numerical results or even diverged results. In nonacoustic applications where large computational domains are not necessary, such as simulations of the near-field of free jet flows, the open boundary conditions in the cross-streamwise directions are also very important because they must allow the entrainment of the ambient fluids (Jiang and Luo 2003). In other words, they must allow the ambient fluids to come into and to go out of the computational domain in a time-dependent simulation. These open boundary conditions also belong to the category of artificial boundaries.

Artificial boundaries such as the open boundaries appear when the computational domain forms only part of the entire flow field in order to reduce the computational costs. They require physically meaningful approximations of the flow and are often difficult to formulate. For the derivation of far-field boundary conditions for compressible flows, all viscous effects are normally neglected, leading to the Euler equations. Although this assumption is in general questionable for vortical flows, it is a reasonable condition under practical aspects such as a large extension of the domain and a highly stretched grid with a coarse resolution in the vicinity of the far-field boundaries. The three-dimensional Euler equations for a compressible flow are a hyperbolic system of five equations with five real eigenvalues (Thompson 1987). These eigenvalues define the directions along which information is transported. Owing to different signs of the eigenvalues, subsonic flows (Mach number $M < 1$) and supersonic flows ($M > 1$) have to be distinguished. If a positive mean flow direction for a supersonic flow is assumed, all eigenvalues are positive. This is the most uncritical case because the transport of information takes place only in a positive coordinate direction. For a flow boundary with the fluids coming into the domain, the values of all variables at the boundary can be prescribed by the undisturbed ambient conditions. For a flow boundary with the fluids going out of the domain, the values of all variables at the boundary can be prescribed by the numerical solutions of the internal domain, using extrapolation of the solutions for the grid points near the boundary.

For a subsonic flow ($M < 1$), four eigenvalues are positive and only one eigenvalue is negative, as shown in Figure 2.1. In contrast to the supersonic case, this means that four variables have to be prescribed by the undisturbed flow at the inlet and one variable has to be extrapolated from the inner region to the boundary. Correspondingly, at the outlet, four variables at the boundary are given by the internal field, whereas one variable is defined from outside. The latter may lead to reflections. In this context, the nonreflecting boundary conditions and the Navier–Stokes characteristic boundary conditions (NSCBC) discussed before can be used to avoid the wave reflections.

For simulations of the near-field of free jet flows, the open boundary conditions in the cross-streamwise directions can also be formulated based upon NSCBC, which allows the entrainment of the ambient fluids. At the boundary, flow variables other than the velocity component normal to the boundary can be prescribed by the physical conditions. The local one-dimensional inviscid (LODI) relations can then be used to provide compatible relations between the physical boundary conditions and the amplitudes of characteristic waves crossing the boundary. Similar to the Euler equations, the Navier–Stokes equations can be transformed into a characteristic form at the boundary. This governing equation for the cross-streamwise velocity component can then be solved to obtain the instantaneous velocity in the direction normal to the boundary. As shown by Jiang and Luo (2003), the open boundary condition formulated upon NSCBC gives a reasonable prediction of the entrainment of the ambient fluids.

B. Periodic and Symmetry Boundary Conditions

Periodic boundaries were already mentioned when inflow boundary conditions were discussed. In practical applications, the applicability of periodic boundary condition is restricted to flow configurations that are indeed periodic owing to their geometry, such as flow around turbine blades, or channel, pipe, and duct flows with one or more statistically homogeneous flow directions. In temporal DNS or LES, where the spectral method (Geurts 2004) is used, a periodic boundary is often encountered. A periodic boundary may be regarded as an overlapping inflow and outflow boundary or an internal face boundary. An internal face boundary or a fluid interior boundary does not represent a challenge because the flow field outside the domain under consideration is a known quantity from computations on the other side of the domain. When flow periodicity is assumed, the flow field outside the computational domain is also a known quantity, provided by the solutions inside the computational domain in the upstream end. Therefore, the specification of periodic boundaries in practical CFD should be straightforward, but care should be taken to ensure that the numerical accuracy of the scheme is not affected at the boundary, which can be done by appropriate programming.

Symmetry boundaries are similar to periodic boundaries, in that the flow field outside the domain is a known quantity. The difference between a symmetry boundary and a periodic boundary is that the flow field is "mirrored" across a symmetry boundary, while the flow repeats itself across a periodic boundary. Symmetry boundaries are encountered when a half or a quarter of the flow domain is considered in the simulation due to the existence of symmetry in the flow field. Examples include axisymmetric simulations of round jets or plumes. Although symmetry does not exist in a fully three-dimensional turbulent flow field—in fact, asymmetric perturbations are needed to break the symmetry of the flow in a three-dimensional DNS or LES with symmetric domain and symmetric physical conditions at the domain boundaries—symmetry boundaries are often used in two-dimensional planar or axisymmetric simulations and some three-dimensional simulations, mainly to reduce the computational costs. The specification of symmetry boundaries in practical CFD is straightforward, but care needs to be taken to ensure that the numerical accuracy of the scheme is not affected at the boundary.

The preservation of the numerical accuracy of the inner computational domain at periodic or symmetry boundaries depends on the details of the numerical methods used (i.e., finite difference, finite volume, or finite element methods), and how the methods are implemented in the computer program. For instance, in a compact finite difference Padé 3/4/6 scheme (Lele 1992), the formal accuracy of sixth order holds in the interior of the computational domain. The scheme is of third-order accuracy at the boundary points, of fourth order next to the boundary points, and of sixth order at inner points only. However, the sixth-order accuracy can be preserved at the symmetry boundary, as shown by Jiang and Luo (2000b) in a simulation of axisymmetric thermal plumes. At the symmetry boundary, which is the jet/plume centerline in an axisymmetric case, the symmetry conditions can be applied to both the primitive variables and their first- and second-order derivatives in the r direction. For a flow variable φ in the governing equation, the flow field is mirrored across the centerline and the variables can be put into two categories:

Category 1: $\varphi(r) = \varphi(-r), \varphi'(r) = -\varphi'(-r), \varphi''(r) = \varphi''(-r)$ (2.15)

Category 2: $\varphi(r) = \varphi(-r), \varphi'(r) = -\varphi'(-r), \varphi''(r) = \varphi''(-r)$ (2.16)

Variables without a sign change across the symmetry boundary belong to the first category given in Equation (2.15), including density ρ and the streamwise velocity component u, while variables that undergo a sign change across the symmetry boundary belong to the second category given in Equation (2.16), notably the radial (cross-streamwise) velocity component v.

The notation $-r$ in these equations refers to the other side of the symmetry boundary. For the Padé 3/4/6 scheme, the expressions here can be used to derive a new set of matrices that preserve the formal sixth-order accuracy of the scheme at the symmetry boundary, which is exemplified as follows.

In the Padé 3/4/6 scheme, the sixth-order accuracy in the inner domain is achieved in the discretization of the first- and second-order derivatives in the governing equations. The first-order derivatives at grid point j with mapped grid distance $\Delta\eta$ (grid mapping occurs when a nonuniform grid is used) can be calculated according to the following discretized equation:

$$\phi'_{j-1} + a\phi'_j + \phi'_{j+1} = b\frac{\phi_{j+1} - \phi_{j-1}}{2\Delta\eta} + c\frac{\phi_{j+2} - \phi_{j-2}}{4\Delta\eta} \qquad (2.17)$$

In Equation (2.17), the constants are given as $a = 3$, $b = (2 + 4a)/3$, and $c = (4 - a)/3$. Equation (2.17) can be applied only at the inner points $3 \le j \le n-2$ with n standing for the total number of grid points in the η direction in the computational domain. At the boundary and next-to-boundary points, the differentiation has to be skewed, leading to the decay of the numerical accuracy from the sixth order in the inner domain. For the Padé 3/4/6 scheme (Lele 1992), a quadratic matrix representing the coefficients on the left-hand side of Equation (2.17), which can be conveniently solved using a tridiagonal matrix solver for the first-order derivatives, is given as

$$\begin{pmatrix} 2 & 4 & & & & & \\ 1 & 4 & 1 & & & & \\ & 1 & 3 & 1 & & & \\ & & \cdots & \cdots & \cdots & & \\ & & & 1 & 3 & 1 & \\ & & & & 1 & 4 & 1 \\ & & & & & 4 & 2 \end{pmatrix} \qquad (2.18)$$

The second-order derivatives at the same point can be calculated according to the following discretized equation:

$$\phi''_{j-1} + a\phi''_j + \phi''_{j+1} = b\frac{\phi_{j+1} - 2\phi_j + \phi_{j-1}}{\Delta\eta^2} + c\frac{\phi_{j+2} - 2\phi_j + \phi_{j-2}}{4\Delta\eta^2} \qquad (2.19)$$

In Equation (2.19), the constants are given as $a = 5.5$, $b = (4a - 4)/3$, and $c = (10 - a)/3$. Equation (2.19) is also applied at the inner points $3 \le j \le n-2$.

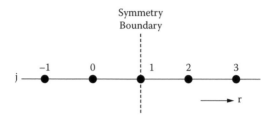

FIGURE 2.3   Schematic of the grid points near a symmetry boundary.

At the boundary and next-to-boundary points, the differentiation has to be skewed, same as the first-order derivatives. For the second-order differentiation of the Padé 3/4/6 scheme, the quadratic matrix representing the coefficients on the left-hand side of Equation (2.19) can be given as

$$
\begin{pmatrix}
1 & 11 & & & & & \\
1 & 10 & 1 & & & & \\
 & 1 & 5.5 & 1 & & & \\
 & & \cdots & \cdots & \cdots & & \\
 & & & 1 & 5.5 & 1 & \\
 & & & & 1 & 10 & 1 \\
 & & & & & 11 & 1
\end{pmatrix}
\qquad (2.20)
$$

For the symmetry boundary condition, the sixth-order formal accuracy of the numerical scheme can be preserved by adjusting the coefficients given in Equations (2.18) and (2.20) using the symmetry conditions given in Equations (2.15) and (2.16). Considering the grid points as shown in Figure 2.3 and categorizing the primitive variables as given in Equations (2.15) and (2.16), for the first-order derivative at $j = 1$, we have

$$\phi_0 = \phi_2, \quad \phi_{-1} = \phi_3, \quad \phi_0' = -\phi_2'$$

Category 1: $\Rightarrow a\phi_1' + 0\phi_2' = 0,$
$$\qquad (2.21)$$

$$\cdots \qquad \cdots \qquad \cdots$$

$$\phi_0 = -\phi_2, \quad \phi_{-1} = -\phi_3, \quad \phi_0' = \phi_2'$$

Category 2: $\Rightarrow a\phi_1' + 2\phi_2' = b\dfrac{\phi_2}{\Delta\eta} + c\dfrac{\phi_3}{2\Delta\eta},$

$$\cdots \qquad \cdots \qquad \cdots$$

$$\qquad (2.22)$$

Following this, the left-hand side of Equation (2.17) for the solution of the first-order derivatives forms a set of discretized tridiagonal equations whose coefficients are given by the following tridiagonal matrices for both categories:

Category 1:
$$\begin{pmatrix} 3 & 0 & & & & & \\ 1 & 3 & 1 & & & & \\ & 1 & 3 & 1 & & & \\ & & \cdots & \cdots & \cdots & & \\ & & & 1 & 3 & 1 & \\ & & & & 1 & 4 & 1 \\ & & & & & 4 & 2 \end{pmatrix} ;$$

(2.23)

Category 2:
$$\begin{pmatrix} 3 & 2 & & & & & \\ 1 & 3 & 1 & & & & \\ & 1 & 3 & 1 & & & \\ & & \cdots & \cdots & \cdots & & \\ & & & 1 & 3 & 1 & \\ & & & & 1 & 4 & 1 \\ & & & & & 4 & 2 \end{pmatrix}$$

Similarly, for the solution of the second-order derivatives as shown in Equation (2.19), by categorizing the primitive variables at $j = 1$, we have

Category 1:

$$\phi_0 = \phi_2, \quad \phi_{-1} = \phi_3, \quad \phi_0'' = \phi_2''$$

$$\Rightarrow a\phi_1'' + 2\phi_2'' = b\frac{2\phi_2 - 2\phi_1}{\Delta\eta^2} + c\frac{2\phi_3 - 2\phi_1}{4\Delta\eta^2},$$

$$\cdots \qquad \cdots \qquad \cdots$$

(2.24)

Category 2:

$$\phi_0 = -\phi_2, \quad \phi_{-1} = -\phi_3, \quad \phi_0'' = -\phi_2''$$

$$\Rightarrow a\phi_1'' + 0\phi_2'' = b\frac{-2\phi_1}{\Delta\eta^2} + c\frac{-2\phi_1}{4\Delta\eta^2},$$

(2.25)

$$\cdots \qquad \cdots \qquad \cdots$$

The corresponding tridiagonal coefficient matrices are defined as

$$
\text{Category 1: }
\begin{pmatrix}
5.5 & 2 & & & & & \\
1 & 5.5 & 1 & & & & \\
 & 1 & 5.5 & 1 & & & \\
 & & \cdots & \cdots & \cdots & & \\
 & & & 1 & 5.5 & 1 & \\
 & & & & 1 & 10 & 1 \\
 & & & & & 11 & 1
\end{pmatrix} ;
$$

$$
\text{Category 2: }
\begin{pmatrix}
5.5 & 0 & & & & & \\
1 & 5.5 & 1 & & & & \\
 & 1 & 5.5 & 1 & & & \\
 & & \cdots & \cdots & \cdots & & \\
 & & & 1 & 5.5 & 1 & \\
 & & & & 1 & 10 & 1 \\
 & & & & & 11 & 1
\end{pmatrix}
$$

$$\tag{2.26}$$

The new set of matrices given in Equations (2.23) and (2.26) can be easily implemented in the computer program to achieve the formal numerical accuracy at the symmetry boundary. The case shown here indicates that the advantageous accuracy offered by the high-order numerical schemes used in DNS and LES can be preserved for periodic and symmetry boundaries.

C. Boundary Conditions: Other Relevant Issues

Discussions on the boundary conditions presented above are mainly made for the primitive variables and their derivatives. In practical CFD applications, there are also situations where boundary conditions need to be specified for fluxes such as mass, momentum, and energy fluxes. In principle, they do not represent special difficulties because they are simply combinations of different flow variables. For example, mass flow inlet can be used in compressible flows to prescribe mass flow rate at the inflow boundary. Boundary condition formulation strategies such as the NSCBC (Poinsot and Lele 1992) can still be used to specify such boundary conditions.

Apart from velocity inflow/outflow, wall, periodic, and symmetry boundaries, pressure boundary may also be used in practical CFD, including pressure inlet boundary and pressure outlet boundary. Normally, a pressure outlet must be used when the problem is set up with a pressure inlet. For time-dependent DNS and LES, the pressure fluctuations at the boundaries also need to be dealt with, in a manner similar to the velocity boundary. For incompressible flows, the pressure inputs at the boundaries define the boundary pressures and the pressure change across the domain. In this case, velocities at the boundary are not fixed and change with the local pressure gradients at the boundary. Other flow quantities at the boundary may be extrapolated from the interior domain. For incompressible flows, pressure boundaries may be used in open boundary domain problems. For compressible flows, a pressure inlet/outlet also implies a temperature condition at the inlet/outlet. Once again, the NSCBC strategy can be used for DNS and LES. In some applications, inlet vent or intake fan and exhaust fan or outlet vent boundary conditions may be encountered, which are closely related to pressure boundaries. An inlet vent or intake fan boundary represents an inlet vent or intake fan with specified loss coefficient or pressure jump, flow direction, and ambient (inlet) pressure and temperature, whereas an exhaust fan or outlet vent represents an external exhaust fan or outlet vent with specified pressure jump or loss coefficient and ambient (discharge) pressure and temperature.

Finally, it is worth noting that the performance of boundary conditions may be controlled by adjusting the computational domain size. For instance, selecting boundary location and shape such that flow goes either in or out of the domain will lead to better performance of an inflow/outflow boundary. The key point here is that boundary conditions are used to approximate the flow field at the boundary; consequently, the closer the boundary condition approximates the physical condition at the boundary, the better the performance will be. In addition, boundaries should not be placed in flow regions where large gradients in the direction normal to the boundary exist. Furthermore, the performance of boundary conditions is also related to the grid or mesh distribution and the numerical methods used in the simulations. The quality of numerical results of DNS or LES is the combined product of the boundary conditions, numerical methods, and grid resolutions used in the simulation. Some applications of boundary conditions in practical DNS and LES will be shown in subsequent chapters (from Chapter 4 onwards).

REFERENCES

Abdallah, S. and Dreyer, J. 1998. Dirichlet and Neumann boundary conditions for the pressure Poisson equation of incompressible flow. *International Journal for Numerical Methods in Fluids* 8: 1029–1036.

Bagwell, T.G., Adrian, R.J., Moser, R.D., and Kim, J. 1993. Improved approximation of wall shear stress boundary conditions for large-eddy simulation. In *Near-Wall Turbulent Flows*, eds. R.M.C. So, C.G. Speziale, and B.E. Launder, 265–275. Amsterdam, The Netherlands: Elsevier.

Balaras, E. and Benocci, C. 1994. Subgrid scale models in finite-difference simulations of complex wall bounded flows. *AGARD CP 551*, 2.1–2.6.

Balaras, E., Benocci, C., and Piomelli, U. 1996. Two-layer approximate boundary conditions for large-eddy simulations. *AIAA Journal* 34: 1111–1119.

Baum, M., Poinsot T., and Thévenin D. 1995. Accurate boundary conditions for multicomponent reactive flows. *Journal of Computational Physics* 116: 247–261.

Breuer, M. 2007. 5.2 Boundary conditions for LES. In *Large-Eddy Simulation for Acoustics*, eds. C. Wagner, T. Hüttl, and P. Sagaut, 201–216. Cambridge, UK: Cambridge University Press.

Cabot, W.H. and Moin, P. 1999. Approximate wall boundary conditions in the large-eddy simulation of high Reynolds number flows. *Flow, Turbulence and Combustion* 63: 269–291.

Danaila, I. and Boersma, B.J. 2000. Direct numerical simulation of bifurcating jets. *Physics of Fluids* 12: 1255–1257.

di Mare, L., Klein, M., Jones, W. P., and Janicka, J. 2006. Synthetic turbulence inflow conditions for large-eddy simulation. *Physics of Fluids* 18: 1–11.

Druault, P., Lardeau, S., Bonnet, J.-P., Coiffet, F., Delville, J., Lamballais, E., Largeau, J.F., and Perret, L. 2004. Generation of three-dimensional turbulent inlet conditions for large-eddy simulation. *AIAA Journal* 42: 447–456.

Geurts, B.J. 2004. *Elements of direct and large-eddy simulation*. Philadelphia PA: Edwards.

Gresho, P.M. and Sani, R.L. 1987. On pressure boundary conditions for the incompressible Navier–Stokes equations. *International Journal for Numerical Methods in Fluids* 7: 1111–1145.

Grötzbach, G. 1981. Numerical simulation of turbulent temperature fluctuations in liquid metals. *International Journal of Heat and Mass Transfer* 24: 475–490.

Grötzbach, G. 1987. Direct numerical and large-eddy simulation of turbulent channel flows. In *Encyclopaedia of Fluid Mechanics*, ed. N.P. Cheremisinoff, 6: 1337–1391, Houston, TX: Gulf Publ.

Grötzbach, G. and Wörner, M. 1999. Direct numerical and large-eddy simulations in nuclear applications. *International Journal of Heat and Mass Transfer* 20: 222–240.

Hinze, J.O. 1975. *Turbulence*, 2nd ed., New York: McGraw-Hill Inc.

Hoffmann, G. and Benocci, C. 1995. Approximate wall boundary conditions for large-eddy simulations. In *Advances in Turbulence V*, ed. R. Benzi, 222–228. Dordrecht, Germany: Kluwer.

Hussain, A.K.M.F. and Zaman, K.B.M.Q. 1981. The preferred mode of the axisymmetric jet. *Journal of Fluid Mechanics* 110: 39–71.

Jarrin, N., Benhamadouche, S., Laurence, D., and Prosser, R. 2006. A synthetic-eddy-method for generating inflow conditions for large-eddy simulations. *International Journal of Heat and Fluid Flow* 27: 585–593.

Jiang, X. and Luo, K.H. 2000a. Combustion-induced buoyancy effects of an axisymmetric reactive plume. *Proceedings of the Combustion Institute* 28: 1989–1995.

Jiang, X. and Luo, K.H. 2000b. Direct numerical simulation of the puffing phenomenon of an axisymmetric thermal plume. *Theoretical and Computational Fluid Dynamics* 14: 55–74.

Jiang, X. and Luo, K.H. 2003. Dynamics and structure of transitional buoyant jet diffusion flames with sidewall effects. *Combustion and Flame* 133: 29–45.

Jiang, X., Siamas, G.A., and Wrobel, L.C. 2008. Analytical equilibrium swirling inflow conditions for computational fluid dynamics. *AIAA Journal* 46: 1015–1018.

Jiang, X., Zhao, H., and Cao, L. 2004. Direct computation of a heated axisymmetric pulsating jet. *Numerical Heat Transfer Part A—Applications* 46: 957–979.

Jiang, X., Zhao, H., and Cao, L. 2006. Numerical simulations of the flow and sound fields of a heated axisymmetric pulsating jet. *Computers & Mathematics with Applications* 51: 643–660.

Jiang, X., Zhao, H., and Luo, K.H. 2007a. Direct computation of perturbed impinging hot jets. *Computers & Fluids* 36: 259–272.

Jiang, X., Zhao, H., and Luo, K. H. 2007b. Direct numerical simulation of a non-premixed impinging jet flame. *ASME Journal of Heat Transfer* 129: 951–957.

Klein, M., Sadiki, A., and Janicka, J. 2003. A digital filter based generation of inflow data for spatially developing direct numerical or large eddy simulations. *Journal of Computational Physics* 186: 652–665.

Lee, S., Lele S., and Moin, P. 1992. Simulation of spatially evolving compressible turbulence and the application of Taylor's hypothesis. *Physics of Fluids A* 4: 1521–1530.

Lele, S.K. 1992. Compact finite difference schemes with spectral-like resolution. *Journal of Computational Physics* 103: 16–42.

Lund, T.S., Wu X., and Squires, K.D. 1998. Generation of turbulent inflow data for spatially-developing boundary layer simulation. *Journal of Computational Physics* 140: 233–258.

Manhart, M. 2001. Analyzing near-wall behaviour in a separating turbulent boundary layer by DNS. In *Direct and Large-Eddy Simulation IV*, eds. B.J. Geurts, R. Friedrich, and O. Métais, 81–88. Dordrecht, Germany: Kluwer.

Mitchell, B.E., Lele, S.K., and Moin, P. 1999. Direct computation of the sound generated by vortex pairing in an axisymmetric jet. *Journal of Fluid Mechanics* 383: 113–142.

Nordström, J., Mattsson, K., and Swanson, C. 2007. Boundary conditions for a divergence free velocity–pressure formulation of the Navier–Stokes equations. *Journal of Computational Physics* 225: 874–890.

Piomelli, U. and Balaras, E. 2002. Wall-layer models for large-eddy simulations. *Annual Review of Fluid Mechanics* 34: 349–374.

Piomelli, U., Ferziger, J.H., Moin, P., and Kim, J. 1989. New approximate boundary conditions for large-eddy simulations of wall-bounded flows. *Physics of Fluids A* 1: 1061–1068.

Poinsot, T. and Veynante, D. 2001. *Theoretical and numerical combustion*. Philadelphia, PA: Edwards.

Poinsot T. and Lele, S.K. 1992. Boundary conditions for direct simulation of compressible viscous flows. *Journal of Computational Physics* 101: 104–129.

Pope, S.B. 2000. *Turbulent Flows*. Cambridge, UK: Cambridge University Press.

Sagaut, P. 2006. *Large eddy simulation for incompressible flows: An introduction*. Berlin: Springer.

Sandham, N.D. 1994. The effect of compressibility on vortex pairing. *Physics of Fluids* 6: 1063–1072.

Sani, R.L., Shen, J., Pironneau, O., and Gresho, P.M. 2006. Pressure boundary condition for the time-dependent incompressible Navier–Stokes equations. *International Journal for Numerical Methods in Fluids* 50: 673–682.

Schumann, U. 1975. Subgrid scale model for finite difference simulations of turbulent flows in plane channels and annuli. *Journal of Computational Physics* 18: 376–404.

Spalart, P.R. 1988. Direct simulation of a turbulent boundary layer up to $Re_\theta = 1410$. *Journal of Fluid Mechanics* 187: 61–98.

Spalart, P.R. 2000. Strategies for turbulence modelling and simulations. *International Journal of Heat and Fluid Flow* 21: 252–263.

Spalart, P.R., Jou, W.H., Strelets, M., and Allmaras, S.R. 1997. Comments on the feasibility of LES for wings and on a hybrid RANS/LES approach. In *Advances in DNS/LES*, eds. C. Liu and Z. Liu, 137–148. Columbus, OH: Greyden Press.

Spalart, P.R. and Leonard, A. 1987. Direct numerical simulation of equilibrium turbulent boundary layers. In *Turbulent Shear Flows 5*, eds. F.J. Durst, B.E. Launder, J.L. Lumley, F.W. Schmidt, and J.H. Whitelaw, 234–252., Berlin, Germany: Springer-Verlag.

Speziale, C.G. 1998. Turbulence modeling for time-dependent RANS and VLES: A review. *AIAA Journal* 36: 173–184.

Sutherland, J.C. and Kennedy C.A. 2003. Improved boundary conditions for viscous, reacting, compressible flows. *Journal of Computational Physics* 191: 502–524.

Thompson, K.W. 1987. Time dependent boundary conditions for hyperbolic systems. *Journal of Computational Physics* 68: 1–24.

Uchiyama, T. 2004. Three-dimensional vortex simulation of bubble dispersion in excited round jet. *Chemical Engineering Science* 59: 1403–1413.

Werner, H. and Wengle, H. 1993. Large-eddy simulation of turbulent flow over and around a cube in a plate channel, In *Selected Papers from the 8th Sym. on Turb. Shear Flows*, eds. F. Durst, R. Friedrich, B.E. Launder, F. Schmidt, U. Schumann, and J.H. Whitelaw, 155–168. Berlin: Springer.

Zhang, H.L., Bachman, C.R., and Fasel, H.F. 2000. Application of a new methodology for simulations of complex turbulent flows. AIAA Paper 2000–2535.

CHAPTER 3

Discrete Time Integration Methods

T IME INTEGRATION IS AN INTEGRAL part of DNS and LES. Both DNS and LES require time-dependent simulations due to the unsteady nature of turbulence. As shown in the governing equations in Chapter 1, temporal or time derivatives are included in those equations. Unsteady or time-dependent flows have a fourth coordinate, time, in addition to the three spatial coordinates, which must be discretized. Temporal derivatives are different from spatial derivatives, in that time derivatives are parabolic like—an event at a given instant affects the flow only in the future—whereas for spatial derivatives, an event at any space location may influence the flow anywhere else. Discretization methods for spatial derivatives will be discussed in subsequent chapters when applications of DNS and LES are described. This chapter discusses temporal integration methods for the time derivative terms in the governing equations. These methods for the governing equations of fluid dynamics are very similar to those applied to initial value problems for ordinary differential equations (ODEs). The basic problem is to find the solution ϕ at a short time interval Δt after the initial point. The solution at $t^1 = t^0 + \Delta t$ can then be used as a new initial condition and the solution can be advanced to $t^2 = t^1 + \Delta t$, $t^3 = t^2 + \Delta t$, ... etc.

For the initial value problems for ODEs, the methods can be either explicit or implicit:

$$\text{Explicit: } \phi^{n+1} = \phi^n + f\left(t^n, \phi^n\right)\Delta t \tag{3.1}$$

$$\text{Implicit: } \phi^{n+1} = \phi^n + f(t^{n+1}, \phi^{n+1})\Delta t \tag{3.2}$$

In an explicit temporal integration method, the solution at the time instant t^{n+1} is decided from the solution and the related functions at the time instant t^n; therefore, the solutions are simply time-marched involving straightforward algebra only. In an implicit method, on the contrary, an equation has to be solved to obtain the solution at the time instant t^{n+1} due to the involvement of functions at the time instant t^{n+1}. Implicit methods are more intensive computationally for each time step compared to explicit methods. However, they do not have stringent limitations in time step for numerical stability, in contrast to explicit methods that normally have time-step restrictions for stability and convergence. Compared with implicit methods, explicit methods may offer better accuracy.

The temporal integration of the governing equations for fluid dynamics follows similar methods to that for the initial value problems for ODEs. For the sake of convenience, the fundamental governing equations for an unsteady, three-dimensional, compressible, and viscous flow, given in Equations (1.1)–(1.5), are written in vector form as

$$\frac{\partial \mathbf{U}}{\partial t} = -\frac{\partial \mathbf{F}}{\partial x} - \frac{\partial \mathbf{G}}{\partial y} - \frac{\partial \mathbf{H}}{\partial z} + \mathbf{J} \tag{3.3}$$

where the column vectors **U**, **F**, **G**, **H**, and **J** are given by

$$\mathbf{U} = \begin{bmatrix} \rho \\ \rho u \\ \rho v \\ \rho w \\ \rho(e + V^2/2) \end{bmatrix}$$

$$\mathbf{F} = \begin{bmatrix} \rho u \\ \rho u^2 + p - \tau_{xx} \\ \rho v u - \tau_{xy} \\ \rho w u - \tau_{xz} \\ \rho(e + V^2/2)u + pu - k\partial T/\partial x - u\tau_{xx} - v\tau_{xy} - w\tau_{xz} \end{bmatrix}$$

$$
G = \begin{bmatrix}
\rho v \\
\rho u v - \tau_{yx} \\
\rho v^2 + p - \tau_{yy} \\
\rho w v - \tau_{yz} \\
\rho(e + V^2/2)v + pv - k\partial T/\partial y - u\tau_{yx} - v\tau_{yy} - w\tau_{yz}
\end{bmatrix}
$$

$$
H = \begin{bmatrix}
\rho w \\
\rho u w - \tau_{zx} \\
\rho v w - \tau_{zy} \\
\rho w^2 + p - \tau_{zz} \\
\rho(e + V^2/2)w + pw - k\partial T/\partial z - u\tau_{zx} - v\tau_{zy} - w\tau_{zz}
\end{bmatrix}
$$

$$
J = \begin{bmatrix}
0 \\
\rho f_x \\
\rho f_y \\
\rho f_z \\
\rho(u f_x + v f_y + w f_z) + \rho \dot{q}
\end{bmatrix}
\tag{3.4}
$$

The temporal integration is essentially finding the variables in the column vector **U**, for a new time instant, where the right-hand side in Equation (3.3) contains the spatial discretization.

For incompressible flows, the time integration needs special attention. With an incompressible solver, the continuity equation as shown in Equation (1.17) cannot be used directly since it does not contain any temporal derivative term. The time integration is performed for Equations (1.18)–(1.21). However, pressure is an unknown quantity in the Navier–Stokes momentum equations and needs to be solved from the pressure Poisson Equation (1.23) at each time step.

For DNS and LES, the time scales of turbulence are very small and the evolution of turbulent structures is very rapid. As a result, the time step in DNS and LES is often not determined by stability considerations, but rather by the time resolution requirements. From a numerical point of view, explicit methods generally provide high accuracy in time, are easy to parallelize, and are computationally more efficient because implicit methods would require much larger memory to solve the discretized equations at each time step. Consequently, the integration of the governing equations in time for DNS is almost exclusively based on explicit methods because of the very large number of mesh points and the high accuracy required. In LES, both explicit and implicit methods have been used. In addition, the temporal integration for DNS and LES very rarely uses numerical methods that are lower than second-order formal accuracy due to the large numerical errors involved in the lower-order time integration; therefore, first-order time integration schemes are not discussed here. In the following sections, Runge–Kutta methods for time integration that belong to the category of one-step methods are presented first, followed by a discussion on the linear multistep methods, including the Adams–Bashforth and Adams–Moulton methods. Finally, other time integration methods include general implicit methods, and other second-order time integration schemes are presented.

I. HIGH-ORDER RUNGE–KUTTA (RK) METHODS

Runge–Kutta (RK) methods are perhaps the most widely used time integration methods for DNS and LES. RK methods are an important class of methods for integrating initial value problems formed by ODEs that encompass a wide selection of numerical methods, and were initially developed by the German mathematicians C.D.T. Runge and M.W. Kutta in the latter half of the nineteenth century. RK methods have been used broadly in DNS and LES in recent years to time-march the system of differential algebraic equations from the discretization of the governing equations. RK methods belong to the category of one-step methods (but with a few substeps or stages) for time integration, which refer only to values at one previous time instant to determine the current values.

Unlike the first-order explicit Euler's method, which has a large truncation error per step as it only evaluates derivatives at the beginning of the interval (i.e., at t^n, in evolving the solution from t^n to t^{n+1}), Runge–Kutta methods make use of several intermediate points of the interval from t^n to t^{n+1}. Higher-order temporal accuracy (higher than second order) may be obtained with the RK methods.

Consider a general governing equation given by

$$U_t = R(U) \tag{3.5}$$

where R represents a functional containing all spatial derivatives and forcing terms such as the one given on the right-hand side of Equation (3.3). Let U^n denote the numerical approximation of U at $t_n = n\Delta t$, where $t_0 = 0$ is the initial time. The basic idea of RK methods is to provide time-marching from the nth time step to $(n + 1)$th time step by building a series of "stages" or "substeps" that approximates the solution U at various points using samples of $R(U)$ from the series of early stages. Finally, the numerical solution U^{n+1} is constructed from a combinations of U^n and all the approximations found at the precomputed stages.

As an example, an explicit fourth-order Runge–Kutta scheme to advance the solutions from a temporal level n to $n + 1$ can be given as

$$U^{(0)} = U^n$$

$$U^{(1)} = U^n + \Delta t\, b_1 R(U^{(0)})$$

$$U^{(2)} = U^n + \Delta t\, b_2 R(U^{(1)}) \tag{3.6}$$

$$U^{(3)} = U^n + \Delta t\, b_3 R(U^{(2)})$$

$$U^{n+1} = U^n + \Delta t \sum_{j=0}^{3} a_j R(U^{(j)})$$

where the coefficients are $a_0 = 1/6$, $a_1 = a_2 = 1/3$, $a_3 = 1/6$; $b_1 = 1/2$, $b_2 = 1/2$, $b_3 = 1$, respectively. The RK4 scheme in Equation (3.6) is the classical fourth-order Runge–Kutta method.

There are a number of different variants of Runge–Kutta methods developed to be more efficient than the classical RK scheme. For the time integration in DNS and LES, explicit high-order Runge–Kutta schemes are computationally efficient and easy to parallelize, and provide high accuracy in time. However, the number of stages in the Runge–Kutta scheme is larger than the order of accuracy if the order of accuracy is higher than four. As a result, such methods require intensive computation. Low-storage Runge–Kutta schemes (Williamson 1980), including third- and fourth-order schemes, which only require two storage locations, have been developed by Carpenter and Kennedy (1994a; 1994b).

As an example of RK3, a family of three-step, compact-storage, third-order Runge–Kutta schemes, derived by Wray (1986), is presented. Two storage locations are employed for each time-dependent variable and at each substep at these locations, say U_1 and U_2, are updated simultaneously as follows:

$$U_1^{new} = c_1 R(U_1^{old})\Delta t + U_2^{old}, \quad U_2^{new} = c_2 R(U_1^{old})\Delta t + U_2^{old} \qquad (3.7)$$

The constants (c_1, c_2) in Equation (3.7) are chosen to be (2/3, 1/4) for substep 1, (5/12, 3/20) for substep 2, and (3/5, 3/5) for substep 3. At the beginning of each full time step, U_1 and U_2 are set to be equal. The data in U_1 is used to compute the right-hand side $R(U_1)$ of Equation (3.3). The computed right-hand side $R(U_1)$ may be stored in U_1 to save storage (overwriting the old U_1). Equation (3.7) is then used to update U_1 and U_2. In Equation (3.7), the time step Δt is limited by the Courant–Friedrichs–Lewy (CFL) condition for stability.

In DNS or LES, a number of requirements need to be complied with when the time integration is performed, including time accuracy, stability, and computational efficiency. The formal order of accuracy gives a good indication of the accuracy of the scheme. For DNS, the time integration schemes used are normally third- or fourth-order accurate, whereas for LES they are from second- to fourth-order accurate. For explicit RK schemes, the formal order of accuracy corresponds to different CFL stability limits. For schemes with accuracy in between second and fourth order, RK methods gain larger stability regions with growing order of accuracy (Drikakis and Rider 2005). For computational efficiency reasons, RK methods with fifth order and above are rarely used because of complexity. In practical simulations of DNS and LES, RK3 and RK4 are widely used because they provide a good compromise between accuracy, stability, and computational efficiency. Apart from the explicit RK methods, there is also a large class of implicit RK methods as described by Ascher and Petzold (1998) for ODEs. However, implicit RK methods are not preferred in DNS and LES, again due to complexity.

Runge–Kutta schemes are mostly suitable for time integration of governing equations with "inexpensive" evaluation of function $R(U)$ in the governing equation $U_t = R(U)$. The evaluation of function $R(U)$ for incompressible flow equations is typically viewed as "expensive" (Drikakis and Rider 2005) because it involves the solutions of the elliptical pressure Poisson equations at each time step. Due to this reason, Runge–Kutta schemes have been mainly used for compressible flow simulations. For compressible flows involving shock waves, total variation diminishing (TVD) is always a preferred property of the numerical schemes because it does not provide nonphysical solutions near the shock waves. The general RK methods given in Equation (3.5) can be made TVD under certain conditions. Gottlieb and Shu (1998) used RK3 that is compatible with TVD, essentially nonoscillatory (ENO) or weighted ENO (WENO) schemes. Their RK method is TVD in the sense that the temporal operator itself does not increase the total variation of the solution. The TVD property of the time integration scheme plays an important role in the time-marching of nonlinear hyperbolic problems. Gottlieb and Shu (1998) also showed that the TVD property can be achieved with a four-stage, fourth-order RK4 method, which is, however, quite intensive computationally and therefore not very useful in practical simulations. For aeroacoustic applications, minimal dissipation and dispersion errors need to be achieved by the numerical methods. By choosing the coefficients of the RK methods, the dissipation and dispersion errors for the propagation of waves can be minimized. The optimized schemes are referred to by Hu et al. (1996) as low-dissipation, low-dispersion RK schemes, which are applicable to different spatial discretization methods and are of low storage.

II. LINEAR MULTISTEP METHODS: ADAMS–BASHFORTH AND ADAMS–MOULTON METHODS

Runge–Kutta methods are one-step methods; that is, only values of the immediate previous time step are used to determine the values at the current time step. Multistep methods refer to the use of function values from several previous time steps in an effort to achieve greater accuracy. However, such a claim needs to be assessed carefully since there might be other deficiencies. Note that RK schemes can conveniently achieve high-order temporal accuracy. In the case of linear multistep methods, a linear combination of the function values from previous time steps is used. Linear multistep methods such as the Adams–Bashforth and Adams–Moulton methods have been applied to DNS and, in particular, LES.

A. The Adams–Bashforth Methods

The Adams–Bashforth methods were developed by John Couch Adams, a British mathematician and astronomer. The methods were then used by Francis Bashforth, to solve a differential equation modeling the capillary action, who became associated with these methods. The second-order

Adams–Bashforth scheme for time integration has been commonly used in LES. It takes the simple form of

$$U^{n+1} = U^n + \frac{1}{2}[3R(U^n) - R(U^{n-1})]\Delta t \qquad (3.8)$$

Apparently the method requires the storage of the function values for two preceding time steps. This can be a problem at the start of the simulation. In a practical simulation, this is handled by the use of lower-order methods, or sometimes Runge–Kutta methods, until sufficient data along the temporal axis has built up. Note that variable time-step sizes may cause difficulties in applying the Adams–Bashforth method. Usually an interpolation over the time interval t^n to t^{n+1} followed by an integration of the interpolated function can be applied; that is,

$$U^{n+1} = U^n + R(U^n)(\Delta t)^n + \frac{(\Delta t)^n}{2(\Delta t)^{n-1}}[R(U^n) - R(U^{n-1})]\Delta t \qquad (3.9)$$

where $(\Delta t)^n = t^{n+1} - t^n$ and $(\Delta t)^{n-1} = t^n - t^{n-1}$. Alternatively, the interpolation can be performed over the previous two time steps, leading to

$$U^{n+1} = U^n + \left(1 + \frac{(\Delta t)^n}{(\Delta t)^{n-1}}\right)R(U^n)(\Delta t)^n - \frac{(\Delta t)^n}{(\Delta t)^{n-1}}R(U^{n-1})\Delta t \qquad (3.10)$$

The third-order Adams–Bashforth method is widely used in DNS and LES and can be written as

$$U^{n+1} = U^n + \frac{1}{12}[23R(U^n) - 16R(U^{n-1}) + 5R(U^{n-2})]\Delta t \qquad (3.11)$$

At the start of the simulation when the function values for preceding time steps are needed but are not available, lower-order methods or Runge–Kutta methods may be used. For the third-order Adams–Bashforth method, variable time-step sizes can also be dealt with by applying the same idea of interpolation as that for the second-order method.

The Adams–Bashforth methods behave differently from the RK methods in terms of stability. Unlike the RK methods, the stability region of the Adams–Bashforth methods decreases with increasing order (Drikakis and Rider, 2005). However, the Adams–Bashforth methods have several advantages over the RK methods for incompressible flow simulations in terms of computational efficiency. For instance, the RK3 scheme requires three substeps with one pressure solution at each substep and a total of three pressure solutions per time step. However, the Adams–Bashforth method involves only one overall pressure solution. Since the solution of the pressure Poisson equation is not a cheap task, it is preferable to use the Adams–Bashforth method rather than the RK method even though the Adams–Bashforth method requires a smaller CFL number for stability.

For compressible flow applications involving shock waves, strong stability preserving (SSP) is preferred for the time integration method. There are linear multistep methods of an SSP type (e.g., Gottlieb et al. 2001; Ruuth and Hundsdorfer 2005). These methods are the generalization of TVD time integration methods, which can achieve high-order formal accuracy as well.

B. The Adams–Moulton Methods

The Adams–Moulton methods are similar to the Adams–Bashforth methods but are based on implicit formulations instead. In these methods, the interpolation includes the current time level. The first-order method is in fact the backward Euler method:

$$U^{n+1} = U^n + R(U^{n+1})\Delta t \tag{3.12}$$

Unlike the explicit Euler scheme, the implicit scheme is extremely stable and is unstable for a small region only.

A second-order Adams–Moulton method is simply the Crank–Nicholson method based on a trapezoidal integration rule, which is very stable for the two variants given by

$$U^{n+1} = U^n + \frac{1}{2}[R(U^n) + R(U^{n+1})]\Delta t \tag{3.13}$$

or

$$U^{n+1} = U^n + R(U^{n+1/2})\Delta t \tag{3.14}$$

where $R(U^{n+1/2})$ is evaluated at $t^{n+1/2} = (t^n + t^{n+1})/2$ in Equation (3.14).

The Adams–Moulton methods also have a third-order scheme. Corresponding to the third-order Adams–Bashforth method given in Equation (3.11), the third-order implicit Adams–Moulton method is given as

$$U^{n+1} = U^n + \frac{1}{12}[5R(U^{n+1}) + 8R(U^n) - R(U^{n-1})]\Delta t \tag{3.15}$$

Unfortunately, the stability of this third-order scheme is not as good as that of the first- and second-order schemes. In fact, the stability region of the Adams–Moulton methods continues to shrink as the order of formal accuracy increases.

Implicit methods usually result in a system of nonlinear equations or linear equations that is not easy to solve in its own right. In practical implementations of an implicit method, predictor–corrector methods are used instead. A predictor–corrector method involves the use of an explicit method to "predict" a time-advanced solution that is substituted into the implicit formula in the "corrector" step. Predictor–corrector methods normally lead to an enhanced stability region over an explicit method. The Adams–Moulton methods were solely due to John Couch Adams. The name of Forest Ray Moulton (who was a U.S. astronomer) became associated with these methods because he realized that they could be used in conjunction with the Adams–Bashforth methods as a predictor–corrector pair.

III. OTHER TIME INTEGRATION METHODS

A. General Implicit Schemes and Backward Differentiation Formulae

The Adams–Moulton methods are one type of implicit method. A large number of different classes of implicit schemes exist and can be written in general as

$$U^{n+1} = \sum_{i=n-k}^{n} a_i U^i + \Delta t \sum_{i=n-k}^{n+1} b_i R(U^i) \tag{3.16}$$

where a_i and b_i are constants with $b_{n+i} \neq 0$. This scheme employs k previous time-step solutions to approximate the solution at the present time step. For convenience in description, Δt is assumed to be a constant in Equation (3.16). For variable time steps, the coefficients a_i and b_i become functions of the time.

All implicit schemes as given in Equation (3.16) give rise to a computational problem for U^{n+1} that can be written in the form

$$U^{n+1} - \Delta t b_{n+1} R(U^{n+1}) = \sum_{i=n-k}^{n} a_i U^i + \Delta t \sum_{i=n-k}^{n} b_i R(U^i) \tag{3.17}$$

The right-hand side of Equation (3.17) depends on the solutions of previous steps only and is therefore known during the time integration. The left-hand side involves the unknown U^{n+1}. Hence, Equation (3.17) represents a large set of coupled nonlinear algebraic equations defining all of the computational mesh points. The very large number of equations and their nonlinearity imply that direct solution methods cannot possibly be used to obtain a solution. Therefore, iterative methods such as approximate Newton iteration (the Newton's method or the Newton–Raphson method) and pseudotime relaxation (Geurts 2004) have to be used to solve Equation (3.17). In general, implicit methods have limited applications in DNS and LES. Nevertheless, they can be useful in certain applications because of their much-enhanced stability over explicit time integration schemes.

Implicit methods being used in the context of finite volume algorithms have been described by Venkatakrishnan and Mavriplis (1996) and Jameson (1991). Note that there is no fundamental difference in the formulation between finite difference or finite volume methods in the spatial derivatives during a temporal integration process. Note also that multigrid methods may be used together with the finite volume algorithms in which the time-marching is treated as a pseudotime variable.

Among the various implicit methods, the backward differentiation formulae (BDF) are considered advantageous over the Adams–Moulton methods. Very often, BDF methods are used for solving stiff systems of equations. These methods have simple forms and much larger stability regions than the Adams–Moulton methods. They are characterized by evaluating the function at the advance time only. The second-order BDF method can be written as

$$U^{n+1} - \frac{4}{3}U^n + \frac{1}{3}U^{n-1} = \frac{2}{3}R(U^{n+1})\Delta t \tag{3.18}$$

The third-order BDF method can be given as

$$U^{n+1} - \frac{18}{11}U^n + \frac{9}{11}U^{n-1} - \frac{2}{11}U^{n-2} = \frac{6}{11}R(U^{n+1})\Delta t \tag{3.19}$$

The stability region of the third-order BDF method is slightly smaller than its second-order counterpart.

B. Other Second-Order Schemes

Time integration methods that are of second-order accuracy are normally considered inadequate for DNS. However, second-order accurate time integration methods are often used in LES and were very popular in CFD.

Methods such as the Lax–Wendroff technique and MacCormack's technique are second-order accurate in both space and time based on explicit finite difference techniques. Implicit time-marching methods such as the approximately factored Beam–Warming algorithm can also be used to achieve second-order accuracy for time integration.

The Lax–Wendroff method (1960) was developed mainly for the time-marching of Euler equations. An important feature of the Lax–Wendroff method is combined time and space differencing. It can be used for the solution of hyperbolic and parabolic partial differential equations. As shown by Anderson (1985), using a Taylor series expansion in time for all the variables, cancelling the first-order error term in the Taylor series expansion, and expressing the derivatives as second-order central differencing, the Lax–Wendroff method allows explicit calculation of the flow field variables at all the grid points at time $t + \Delta t$ from the known flow field variables at time t.

There are a number of variants of the Lax-Wendroff method. Richtmyer (1963) derived a useful variant in a predictor–corrector format, in which an intermediate time level $n + 1/2$ was involved. Perhaps the most well-known variant of the Lax–Wendroff method is that due to MacCormack (1969). In MacCormack's technique, a backward differencing is used as the predictor and a forward differencing is used as the corrector, or vice versa. Higher-order methods (higher than second order) of the Lax–Wendroff methods are possible. However, the complexity of the algorithms increases geometrically as the order increases. Therefore, these higher-order methods are of very limited use for practical CFD calculations. The Lax–Wendroff methods can also be used in conjunction with the WENO scheme for compressible flow simulations involving shock waves (e.g., Qiu and Shu 2003). The principles of the Lax–Wendroff method can in general be applied to incompressible flows. However, the pressure Poisson equation needs to be solved at each substep, which makes the method costly.

For second-order time integration methods, implicit methods can offer improved stability over the explicit methods. For example, the implicit Beam–Warming methods (Beam and Warming, 1976) are very stable. The methods can be second-order accurate both in space and in time. Ekaterinaris (1999; 2000) showed fourth-order formal accuracy for the implicit operators with higher-order discretization of the right-hand side. In general, the complexity of high-order implicit methods is higher than that of high-order explicit methods. Implicit methods can be computationally less effective than explicit methods because the solution of a set of nonlinear algebraic equations is involved in each step of the time-marching.

REFERENCES

Anderson, John D. 1995. *Computational fluid dynamics: The basics with applications*. New York: McGraw-Hill.

Ascher, U.M. and Petzold, L.R. 1998. *Computer methods for ordinary differential equations and differential-algebraic equations*. Philadelphia PA: SIAM.

Beam, R.M. and Warming, R.F. 1976. An implicit finite difference algorithm for hyperbolic systems in conservation-law form. *Journal of Computational Physics* 22: 87–110.

Carpenter, M.H. and Kennedy, C.A. 1994a. Third-order 2N-storage Runge–Kutta schemes with error control. NASA-TM-109111. NASA Langley Research Center.

Carpenter, M.H. and Kennedy, C.A. 1994b. Fourth-order 2N-storage Runge–Kutta schemes. NASA-TM-109112. NASA Langley Research Center.

Drikakis, D. and Rider, W. 2005. *High-resolution methods for incompressible and Low-Speed flows*. Berlin: Springer.

Ekaterinaris, J.A. 1999. Implicit, high-resolution, compact schemes for gas dynamics and aeroacoustics. *Journal of Computational Physics* 156: 272–299.

Ekaterinaris, J.A. 2000. Implicit high-order accurate-in-space algorithms for the Navier–Stokes equations. *AIAA Journal* 38: 1594–602.

Geurts, Bernard J. 2004. *Elements of direct and large-eddy simulation*. Philadelphia, PA: Edwards.

Gottlieb, S. and Shu, C.W. 1998. Total variation diminishing Runge–Kutta schemes. *Mathematics of Computation* 67: 73–85.

Gottlieb, S., Shu, C.W., and Tadmor, E. 2001. Strong stability-preserving high-order time discretization methods. *SIAM Review* 43: 89–112.

Hu, F.Q., Hussaini, M.Y., and Manthey, J.L. 1996. Low-dissipation and low-dispersion Runge–Kutta schemes for computational acoustics. *Journal of Computational Physics* 124: 177–191.

Jameson, A. 1991. Time dependent calculations using multigrid, with applications to unsteady flows past airfoils and wings, AIAA Paper 91-1596, Reno, NV.

Lax, P.D. and Wendroff, B. 1960. Systems of conservation laws. *Communications on Pure and Applied Mathematics* 13: 2217–237.

MacCormack, R.W. 1969. The effect of viscosity in hypervelocity impact cratering. AIAA Paper 69-354, Cincinnati, OH.

Qiu, J.X. and Shu, C.W. 2003. Finite difference WENO schemes with Lax–Wendroff-type time discretizations. *SIAM Journal on Scientific Computing* 24: 2185–2198.

Richtmyer, R.D. 1963. A survey of difference methods for non-steady fluid dynamics. Technical Report NCAR 63-2, NCAR.

Ruuth, S.J. and Hundsdorfer, W. 2005. High-order linear multistep methods with general monotonicity and boundedness properties. *Journal of Computational Physics* 209: 226–248.

Venkatakrishnan, V. and Mavriplis, D.J. 1996. Implicit method for the computation of unsteady flows on unstructured grids. *Journal of Computational Physics* 127: 380–397.

Williamson, J.H. 1980. Low-storage Runge–Kutta schemes. *Journal of Computational Physics* 35: 48–56.

Wray, A.A. 1986. Very low storage time-advancement schemes. Internal Report, NASA-Ames Research Center, Moffett Field, CA.

CHAPTER 4

DNS of Incompressible Flows

$\mathbf{A}$S DISCUSSED IN CHAPTER 1, incompressible flows represent a broad range of fluid flows with relatively low speeds, typically for Mach number $M < 0.3$. Due to the very high computational costs associated with DNS, state-of-the-art DNS is mainly restricted to low Reynolds number flows in simple geometries. There has been a substantial amount of DNS of incompressible flows in simple geometries due to the relatively low speeds of incompressible flows, and hence relatively lower computational costs compared with high-speed flows when all the scales need to be resolved. One example is DNS of channel flows, which are of great importance to many practical applications and fundamental fluid dynamics research. This chapter presents a few applications of DNS to incompressible flows with some sample results of DNS of channel flows. Discussions on the numerical features of the simulations are also presented.

I. SAMPLE RESULTS: DNS OF CHANNEL FLOWS

A. DNS of a Separated Channel Flow with a Smooth Profile (Marquillie et al. 2008)

Marquillie et al. (2008) solved the incompressible three-dimensional, time-dependent Navier–Stokes equations along with a Poisson equation for calculating pressure for a flow in a channel with a smooth profile. For the spatial discretization, fourth-order central finite differences were used for the second derivatives in the streamwise x direction. All first derivatives of the flow quantities appearing explicitly in the time-advancing scheme and the first derivatives in the x direction were discretized using eighth-order finite differences. Chebyshev collocation was used in the

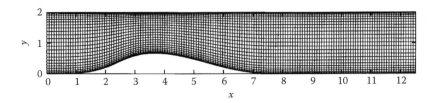

FIGURE 4.1 Mesh distribution: DNS of a separated channel flow with a smooth profile. (Marquillie et al. 2008; with permission from Taylor & Francis.)

wall-normal y direction. The transverse direction z was assumed periodic and was discretized using a spectral Fourier expansion. For time integration, implicit second-order backward Euler differencing was used; the Cartesian part of the Laplacian was taken implicitly, whereas an explicit second-order Adams–Bashforth scheme was used for other operators as well as for the nonlinear convective terms. The three-dimensional system was uncoupled into two-dimensional subsystems and the resulting two-dimensional Poisson equations were solved efficiently using the matrix diagonalization technique. The simulation used a significant amount of computing resources. A spatial resolution of $1536 \times 257 \times 384$ was finally used, which corresponds to a maximum mesh size of 3.9η in the normal direction and 6.8η in the other two directions, where η is the local isotropic Kolmogorov scale. The cross-sectional computational grid is sketched in Figure 4.1.

Several profiles of the mean streamwise velocity are plotted in Figure 4.2. The mean velocity profile at the outlet does not recover the same shape as that at the inlet. In order to study the global motion of the turbulent

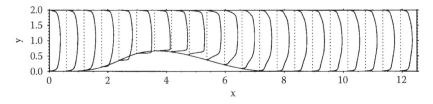

FIGURE 4.2 Profiles of mean streamwise velocity: DNS of a separated channel flow with a smooth profile (Marquillie et al. 2008; with permission from Taylor & Francis.)

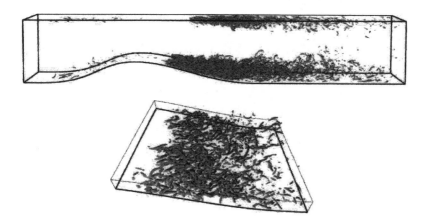

FIGURE 4.3   Isovalues of the second invariant of the velocity gradient tensor: DNS of a separated channel flow with a smooth profile. (Marquillie et al. 2008; with permission from Taylor & Francis.)

structures, isovalues of the second invariant of the velocity gradient tensor indicating the flow vortical structures are presented in Figure 4.3. The vortices generated near the separation are much more intense than the coherent structures generated in a plane channel flow. Consequently, the isovalue used to detect the vortices downstream of the bump was not adapted to detect structures at the inlet of the simulation. The generation of intense coherent structures is nearly steady in time and space, but the localization is different at the two walls and is slightly downstream of the position of the first inflexion point of the mean velocity profile. However, the typical size of the vortices seems comparable at both walls and varies very slowly in time. This behavior may be due to the detection criteria, which only capture well-formed and intense vortices. Almost no vortices are detected in the upstream part, but strong ones are generated close to the wall in the downstream part of the separation region. These vortices interact with those convected in the outer boundary layer. This leads to nontrivial motions of these small near-wall vortices. Only a small part of them is convected by the mean reversal flow very close to the wall; most of them are either destroyed by the shear or by interaction with larger vortices. As shown by the instantaneous velocity field (Figure 4.4), some vortices lift up and occasionally generate an ejection of fluid in the outer region. This study revealed the fine details of the near-wall structures in a separated channel flow using highly accurate numerical methods, which is difficult to achieve using lower-order numerical schemes.

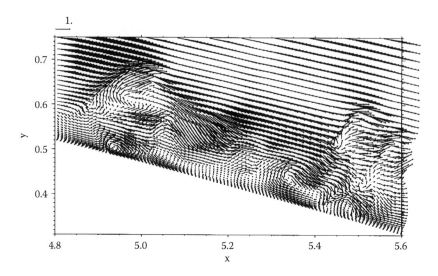

FIGURE 4.4   Velocity vectors within the separation region: DNS of a separated channel flow with a smooth profile. (Marquillie et al. 2008; with permission from Taylor & Francis.)

B. DNS of Turbulent Heat Transfer in Pipe Flows (Redjem-Saad et al. 2007)

The problem of heat transfer in pipe flows is of importance in mechanical and many other engineering fields, and is encountered in a variety of applications such as in heat pipes, combustion chambers, and nuclear reactors. Redjem-Saad et al. (2007) performed a direct numerical simulation of turbulent heat transfer in a pipe with a Reynolds number Re = 5500 based on the pipe radius. The effect of Prandtl number on the flow field was investigated. The flow configuration is a forced, fully developed, incompressible pipe flow of a Newtonian fluid, heated with a uniform heat flux imposed on the pipe wall. The governing equations were discretized on a staggered grid in cylindrical coordinates. Numerical integration was performed using a finite difference scheme, second-order accurate in space and time. The time-advancement employed a fractional step method. A third-order Runge–Kutta explicit scheme and a Crank–Nicolson implicit scheme were used to evaluate the convective and diffusive terms, respectively. Uniform computational grid and periodic boundary conditions were applied to the circumferential and axial directions while the wall temperature fluctuations were assumed to be zero, corresponding to a

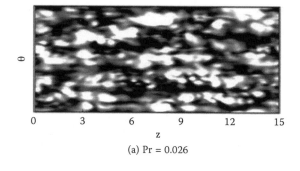

(a) Pr = 0.026

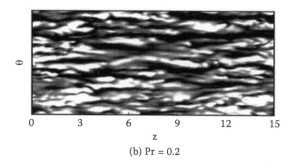

(b) Pr = 0.2

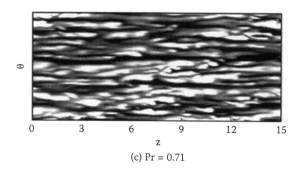

(c) Pr = 0.71

FIGURE 4.5 Instantaneous temperature fluctuations at various Prandtl numbers: DNS of turbulent heat transfer in pipe flows. (Redjem-Saad et al. 2007; with permission from Elsevier Science Ltd.)

mixed-type boundary condition. In this case the time-averaged wall heat flux is uniform in space, and the wall temperature is not time dependent and varies linearly along the streamwise direction.

Figure 4.5 shows the instantaneous temperature fluctuations for three different Prandtl numbers. For Pr = 0.026, a regular distribution is observed while for $Pr = 0.2$, streaky structures are observed. At the higher Pr number, the conductive region becomes thinner, leading to a reduction in molecular heat flux and an enhancement in the turbulent heat flux normal to the wall. These features are more pronounced in Figure 4.5(c), where the Prandtl number is even higher. The Prandtl number is an important parameter in heat transfer problems and is a dimensionless number approximating the ratio of momentum diffusivity (kinematic viscosity) and thermal diffusivity. In heat transfer problems, the Prandtl number controls the thickness of the thermal boundary layer relative to that of the momentum boundary layer. When Pr is small, it means that the heat diffuses very quickly compared to the velocity (momentum). The results indicate that the Prandtl number has a significant impact on the instantaneous temperature fields, and therefore a significant impact on the heat transfer characteristics of turbulent pipe flows.

C. DNS of Turbulent Channel Flow under Stable Stratification (Iida et al. 2002)

Iida et al. (2002) studied a turbulent channel flow under stable stratification by means of DNS. The governing equations solved are the standard set of hydrodynamic equations (Navier–Stokes) with the assumption of the constant physical properties and the Boussinesq approximation. For spatial differentiation, a spectral method was used to obtain the solutions with Fourier series in two directions with periodic boundary conditions. A Chebyshev polynomial expansion was used in the wall-normal direction with a no-slip boundary condition assumed at the two walls, which were kept at different but uniform temperatures.

Figure 4.6 compares the isosurfaces of the nondimensional temperature under neutral (Grashof number $Gr = 0$) and stable density stratification ($Gr = 10^7$), respectively. The Grashof number is a dimensionless number frequently used in buoyant fluid flows and heat transfer problems in enclosures that approximates the ratio of the buoyancy to viscous force acting on a fluid. A significant difference is observed between these two cases with and without buoyancy effect. In the case without buoyancy, the temperature isosurface shown in Figure 4.6(a) is violently torn up, indicating that the temperature is effectively mixed by turbulent motions. In the buoyant case shown in Figure 4.6(b), under stable stratification, undulations of the isosurface become smaller, and thermal mixing due to turbulence is significantly less.

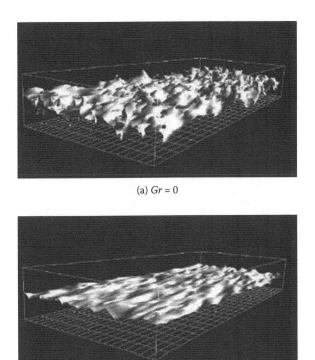

(a) $Gr = 0$

(b) $Gr = 10^7$

FIGURE 4.6   Isosurfaces of nondimensional temperature: DNS of turbulent channel flow under stable stratification. (Iida et al. 2002; with permission from Elsevier Science Ltd.)

Figures 4.7 and 4.8 show the instantaneous flow structures for the velocity vector and the isosurfaces of the pressure fluctuations. The locations of the $x - y$ and $y - z$ planes included in Figure 4.7 are indicated by the black bars in Figure 4.8. A wave-like motion associated with the pressure fluctuation is clearly observed in the central region of the channel. It is noted that a vertically elongated low-pressure region is often generated between the wall and the crest of the wave motions, indicating the close association between them. The streamwise vortices reach down near the wall and are associated with the wave crests. These low-pressure regions are elongated in the horizontal direction. The results indicated the complex flow structures in the flow field and the instantaneous correspondence between velocity and pressure.

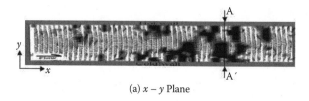

(a) $x - y$ Plane

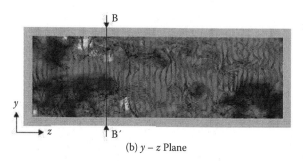

(b) $y - z$ Plane

FIGURE 4.7   Instantaneous velocity vectors in different cross-sections of the case with DNS of turbulent channel flow under stable stratification. (Iida et al. 2002; with permission from Elsevier Science Ltd.)

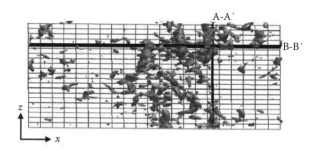

FIGURE 4.8   Instantaneous isosurfaces of the low pressure region of the case with DNS of turbulent channel flow under stable stratification. (Iida et al. 2002; with permission from Elsevier Science Ltd.)

D. DNS of Coherent Structure in Turbulent Open-Channel Flows with Heat Transfer (Yamamoto et al. 2000)

In the DNS study of the open-channel flow with heat transfer by Yamamoto et al. (2000), the governing equations of the incompressible Navier–Stokes equations with the Boussinesq approximation were solved. Numerical integration of the governing equations was based on a fractional step method and time integration was performed using a

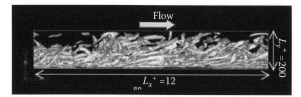

(a) Side View

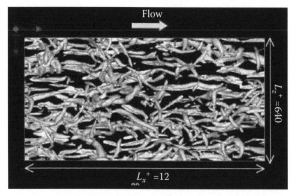

(b) Top View

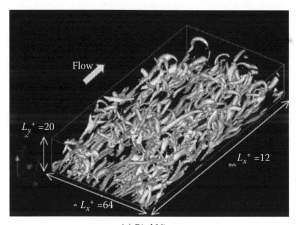

(c) Bird View

FIGURE 4.9 Isosurfaces of second invariant velocity gradient tensor: DNS of the open-channel flow with heat transfer. (Yamamoto et al. 2000; with permission from Springer-Verlag.)

second-order Adams–Bashforth scheme. A second-order central differencing scheme was adopted for the spatial discretization. As the boundary conditions for fluid motion, free-slip condition at the free surface, no-slip condition at the bottom wall and the cyclic conditions in the stream- and the spanwise directions were imposed, respectively. As for the equation of energy, temperatures at the free surface and the bottom wall were kept constant.

Figure 4.9 shows the isosurface representation of a second invariant velocity gradient tensor. The isosurface regions correspond to the strong vorticity containing regions. Near the bottom wall, the streamwise vortex stretched out in the streamwise direction can be seen. This indicates that turbulence is generated near the wall. However, the free surface makes no contribution to turbulence generation in open-channel flows at the relatively low Reynolds number of 200 considered (based on the friction velocity and flow depth).

II. NUMERICAL FEATURES: DNS OF INCOMPRESSIBLE FLOWS

From the above examples, it can be seen that there are some common features in the numerical methods used in the DNS studies of incompressible channel flows. For DNS applications, high-order schemes are normally needed for spatial discretization. In general, lower-order numerical schemes use fewer data points (either in space or time) since smaller stencils are normally needed, and thus require less computing resources, obtaining better speed but potentially lower accuracy than high-order numerical schemes that use more data points and require more computing resources. For the spatial discretization, finite difference schemes with at least second-order numerical accuracy have been used. The spectral method is also used in flow directions when periodic boundary conditions are employed. In addition, Chebyshev collocation or polynomial expansion has been used. These methods will be briefly discussed, including their formulations, advantages, and disadvantages. For the temporal integration of the governing equations, the common feature in these applications is the use of the fractional step method for the time advancement of the incompressible flow governing equations, apart from the time integration methods discussed in Chapter 3. For incompressible flows, the pressure needs to be obtained from the Poisson equation solver, which will be briefly discussed, together with the fractional step method.

Discussions on the boundary conditions were presented in Chapter 2 and were omitted here.

A. Spatial Discretization Schemes

All DNS codes are dependent on the types of numerical schemes used to translate the Navier–Stokes equations into a finite form appropriate for the computational grids/cells. In the process of discretization and numerical solution of the discretized equations, there are competing needs of accuracy, stability, and computing speed. It has been a consensus that high-order numerical schemes should be used in DNS. Highly accurate numerical schemes are needed in DNS because turbulence may not be resolved using low-order numerical schemes such as the first-order scheme, where numerical diffusion can be larger than small-scale turbulent transportation. It is well known that high-order numerical schemes in DNS and LES can significantly improve predictions of vortical and other complex unsteady flows. When accuracy is a crucial requirement of the simulation such as in a DNS, application of high-order schemes is expected to decrease the computing costs since the error in the functional approximation is proportional to h^n for an nth-order accurate numerical scheme with h representing the grid spacing. In a high-order scheme, the error in the functional approximation decreases much faster with the grid spacing than that in a lower-order scheme. In a CFD simulation, the computing time depends on the order of accuracy, the complexity of the method, and the grid resolution. In DNS and LES, to achieve the required level of accuracy, the employment of high-order numerical schemes may reduce the overall computing costs compared with using lower-order numerical schemes, provided that the formulations of the high-order numerical schemes are not overly complex; for example, a small or compact stencil can be used. As discussed by Ekaterinaris (2005), high-order numerical schemes are a broad concept that can include high-order finite difference and finite volume methods, the discontinuous Galerkin method, and spectral and spectral volume methods. Although second-order accuracy schemes have been used in DNS, such as in the channel flow examples by Redjem-Saad et al. (2007) and Yamamoto et al. (2000), most of the current DNS codes for computational fluids and combustion employ numerical methods that are fourth-order accurate or higher.

The majority of the existing DNS codes employ high-order finite difference schemes, due to the fact that their computing costs are generally lower than high-order finite volume methods (Ekaterinaris 2005). In

DNS codes (and in LES codes as well), it is preferable to have methods that are nondissipative. Numerical schemes with dissipations such as the first-order, third-order, and fifth-order schemes normally lead to errors in flow regions with large gradients, which is not suitable for predictions of vortical flow fields. Although dissipative upwind schemes are more stable and easier to converge than nondissipative central differencing schemes, upwind biased high-order schemes (such as the third-order, fifth-order, etc.) inherently introduce some form of artificial smoothing that makes them inappropriate for long time integrations encountered in direct and large-eddy simulations. Consequently, nondissipative central differencing schemes have been predominantly used in DNS codes, where the second-order scheme has been adopted as the numerical scheme with the lowest acceptable accuracy for DNS applications.

High-order central-difference schemes can be conveniently obtained from Taylor series expansion. The stencils for the fourth-, sixth-, and eighth-order accurate symmetric, explicit, centered schemes are five-, seven- and nine-point wide, respectively. Due to the high comput-ing costs involved in the schemes higher than fourth order, they are rarely used in CFD codes. For efficient computing, it is essential for higher-order schemes to use narrow stencils. The compact finite dif-ference (Padé) scheme presented by Lele (1992) was perhaps the first systematic attempt to develop high-order accurate, narrow stencil, finite difference schemes appropriate for problems with a broad range of scales such as those encountered in turbulent flows and combustion. The Padé scheme has been widely used and now is the state-of-the-art numerical method in DNS codes for flow and combustion phenom-ena. The main advantages of the Padé scheme are the low computing costs associated with the small stencil support and the simplicity in boundary condition treatment. For instance, a seven-point wide sten-cil can yield a compact tenth-order accurate scheme. By varying the coefficients in the compact approximations for the first- and second-order derivatives of the Navier–Stokes equations in the computational domain, schemes of different accuracy can be achieved. Lele (1992) systematically presented the Padé schemes with different orders of accuracy, including the fourth-order explicit scheme, the compact fourth-order scheme, sixth-order scheme, eighth-order scheme with tridiagonal matrix inversion, eighth-order scheme with pentadiago-nal matrix inversion, and the tenth-order schemes, which have been broadly used in DNS codes. The details of these schemes can be found in Lele (1992). The formulation of one such scheme will also be presented in the next chapter.

The most significant advantage offered by finite difference schemes is the flexibility in the specification of boundary conditions, which are much more difficult in other accurate methods such as spectral methods. Spectral methods can be used to calculate the spatial derivatives in the governing equations with simple boundary conditions such as periodic boundaries. Although spectral methods are not easy to apply for practical boundaries, complex domains, and compressible flows with discontinuities, they are numerically accurate with low computing costs. Due to this advantage, they have been broadly used in DNS of incompressible flows such as in the channel flow simulations performed by Iida et al. (2002) and Marquillie et al. (2008). These DNS of fluid flows with at least one periodic boundary using spectral methods are often referred to as temporal DNS, in contrast to spatial DNS where no periodicity is assumed.

In finite difference, finite volume, or even in finite element methods, approximate derivatives are obtained by using local information of the solution. In other words, only values of the solution in the neighborhood of the point where the derivative is evaluated are taken into consideration. They are methods based on local information, and therefore the accuracy of the numerical scheme depends on how many and how the neighboring points have been taken into consideration. In contrast to these "local" methods, the spectral method or pseudospectral method (Canuto et al. 1988) is based on a global coupling, which is an implicit differencing method in view of the requirement of solving a linear system of equations. The starting point of the spectral method is the representation of the solution in a set of basic functions that are defined in the entire flow domain. A basic spectral method involves three main steps:

- Step 1. Choose the basis functions: An appropriate set of global basis functions $\phi_n(x)$ for $n = 0, 1, \ldots, N$ has to be selected in order to arrive at a spectral discretization of a specific problem. In a basic spectral method dealing with periodic boundaries, Fourier modes are normally selected. The requirements posed by the boundary conditions decide the choice of basis functions, and ideally, basis functions that individually satisfy boundary conditions are needed.

- Step 2. Expand the desired solution: The flow solution variables, for example, $u(x)$, can be approximated by a finite dimensional expansion in terms of the basis functions

$$u(x,t) \approx u_N(x,t) = \sum_{n=0}^{N} a_n(t)\phi_n(x) \qquad (4.1)$$

In Equation (4.1), $a_n(t)$ represents the time-dependent expansion coefficients.

- Step 3. Determine the expansion coefficients: A weighed residual can be formulated in order to obtain equations from which the coefficients follow. The basic problem may be expressed as $\mathbb{N}(u) = f$ where $\mathbb{N}$ is the total differential operator and f is a possible forcing term. The residual can then be defined as $R(u) = \mathbb{N}(u) - f$. In a practical simulation, the weighed residual approach in terms of $R(u_N)$ is based on

$$\int_{\Omega} W_m(x)R(u_N)dx = 0; \quad m = 0,1,\dots,N \qquad (4.2)$$

In Equation (4.2), Ω represents the flow domain and a set of weighing functions W_m are introduced that measure the deviation of the approximate solution u_N from the analytical solution u. Substituting Equation (4.1) into Equation (4.2) yields a system of differential equations for the unknowns $\{a_n\}$ that need to be solved in order to obtain the flow solution. There are two common choices for the weighing functions. One choice is the Galerkin method, which requires the residual $R(u_N)$ to be orthogonal to the base functions:

$$\int_{\Omega} W(x)\phi_m(x)\phi_n(x)dx = \delta_{mn} \qquad (4.3)$$

In Equation (4.3), δ is the Kronecker delta. In this choice, the weighing function is given as $W_m = W(x)\phi_m(x)$. In another choice of the weighing function, the values of the residual are required to be zero in a set of locations in the flow domain

$$R(u_N)(x_m) = 0; \quad m = 0,1,\dots,N \qquad (4.4)$$

In the latter choice leading to Equation (4.4), the weighing function is given as $W_m = \delta(x - x_m)$, where x_m is the collocation or grid points.

For spectral methods dealing with periodic boundaries, the central objective is to determine the expansion coefficients $\{a_n\}$. The entire simulation can be performed in the spectral space and a transform back to the physical space is carried out only when it is necessary. For spectral methods in a broad sense, the pseudospectral method is a slight adaption of the method described above. In a pseudospectral approach, the basic setting of the method is in the physical space and the grid values of the variable $u_N(x_j)$ need to be approximated instead of the expansion coefficients. In order to obtain derivatives, a transform to the spectral space needs to be applied, leading to the derivatives in the spectral space that can be transformed back to the physical space. When periodic boundary conditions are used, a finite Fourier series may be adopted:

$$u(\mathbf{x},t) = \sum_{k} \hat{u}_{k}(t) e^{i\mathbf{k}\cdot\mathbf{x}} \tag{4.5}$$

In Equation (4.5), $\mathbf{k} = (k_x, k_y, k_z)$ and the summation over k_α runs from $-N_\alpha/2$ to $N_\alpha/2$. The first- and second-order derivatives can be found by multiplication of the Fourier coefficient $\hat{u}_k$ with (ik_α) and $(-k_\alpha^2)$ and transforming the spectral space representation back to the physical space (Canuto et al. 1988). The pseudospectral approach represents an effective method in dealing with the nonlinear terms. In periodic problems, the solution is expanded in a finite number of Fourier modes. Within this finite dimensional subspace, the first and second derivatives of the modes can be represented exactly. There is no discretization error involved and therefore spectral methods are normally very accurate. The only errors that arise in spectral methods are due to the finite number of modes considered, which are the so-called aliasing errors. The aliasing errors can be tackled in various ways and further details can be found in the book by Canuto et al. (1988). When the grid resolution is very high such as in most of the DNS studies, the errors are not significant because the contributions of the smallest scales may be sufficiently small. However, the aliasing errors need to be treated more precisely when the grid resolution is high, such as in some LES applications.

The spectral methods discussed so far have been restricted to periodic boundaries, which have limited practical applications although they are highly accurate. For nonperiodic problems, the use of Fourier expansions is very difficult or not optimal in view of an efficient compliance with boundary conditions. For the choice of the basis functions (step 1 of a basic spectral approach), Fourier modes can no longer be selected for nonperiodic boundaries. Instead, a set of orthogonal polynomials may be used for nonperiodic boundaries, among which the use of Chebyshev polynomials is most popular. For instance, in the examples shown in the previous section (Iida et al. 2002; Marquillie et al. 2008), Chebyshev collocation or Chebyshev polynomial expansion has been used for spatial discretization. In the following, spectral methods for nonperiodic problems are briefly described with focus given to the Chebyshev polynomials. The description closely follows Geurts (2004), which was adapted from the notes by Dijkstra (1999: Pseudospectral collocation methods. Lecture notes JMBC course Computational Fluid Dynamics III, University of Twente, The Netherlands).

The most important procedure in establishing spectral methods for nonperiodic problems is the specification of the collocation derivative matrix, which proceeds in a few steps (Geurts 2004). For general polynomials on general grids, considering the interpolation of a solution u, which is known by its values u_j on grid points $x_j \in [a,b]$, the polynomial of degree $(N+1)$ that vanishes at the grid points can be given as

$$Q(x) = A \prod_{k=0}^{N} (x - x_k) \tag{4.6}$$

In Equation (4.6), $A \neq 0$. Considering $Q'_j = Q'(x_j)$, a cardinal polynomial K_j can be defined as

$$K_j(x) = \frac{Q(x)}{(x - x_j)Q'_j} \tag{4.7}$$

The cardinal polynomial K_j has degree N and satisfies $K_j(x_i) = \delta_{ij}$. The cardinal polynomials may be used to represent the polynomial $(I_N u)$ of degree $\leq N$, which interpolates u at the grid as follows:

$$(I_N u)(x) = \sum_{j=0}^{N} K_j(x) u_j \tag{4.8}$$

It can be verified that $(I_N u)(x_k) = u_k$. Considering $\xi = \xi(x) \in [a,b]$, the quality of the interpolation may be expressed by

$$u(x) = (I_N u)(x) + \frac{u^{(N+1)}(\xi)Q(x)}{(N+1)!A} \tag{4.9}$$

The derivative of the interpolating polynomial at $x = x_i$ is given by

$$\frac{d}{dx}(I_N u)(x_i) = \sum_{j=0}^{N} K'_j(x_i)u_j; \quad i = 0,1,\ldots,N \tag{4.10}$$

Equation (4.10) implies in matrix notation $(I_N u)' = Du$, where D denotes the $(N+1) \times (N+1)$ collocation derivative matrix with elements $D = (d_{ij}) = [K'_j(x_i)]$. The differentiation of K_j can be given as

$$K'_j(x) = \frac{Q'(x)}{(x - x_j)Q'_j} - \frac{Q(x)}{(x - x_j)^2 Q'_j} \tag{4.11}$$

Substituting $x = x_i$ into Equation (4.11), the elements of D can be given as

$$d_{ij} = K'_j(x_i) = \frac{Q'_i(x)}{(x_i - x_j)Q'_j}; \quad i \neq j \tag{4.12}$$

$$d_{jj} = K'_j(x_j) = \frac{Q''_j}{2Q'_j} \tag{4.13}$$

Higher-order derivatives may also be obtained using the collocation derivative matrix D. It can be shown (Geurts 2004) that the second derivative matrix can be obtained by squaring the first derivative matrix $D^{(2)} = D^2$, which can also be generalized as $D^{(m)} = D^m$ $(m \geq 2)$ for a m-order derivative.

The use of Chebyshev polynomials is by far the most popular when the flow domain is nonperiodic, due to the fact that the interpolation error in Equation (4.9) is minimal on a proper grid. The Chebyshev polynomials may be defined by the recursion

$$T_0(x) = 1; \quad T_1(x) = x; \quad T_{n+1}(x) = 2xT_n(x) - T_{n-1}(x); \quad n = 1,2,\ldots \tag{4.14}$$

From the trigonometric representation $T_n[\cos(\theta)] = \cos(n\theta)$, for a $(N + 1)$ th Chebyshev polynomial $T_{N+1}(x)$, it can be obtained that

$$T_{N+1}(x) = 0 \text{ at } x = x_k = \cos\left(\frac{k + \frac{1}{2}}{N+1}\pi\right); \quad k = 0,1,\ldots,N \tag{4.15}$$

The roots given in Equation (4.15) are all simple and in the interior of $(-1, 1)$, and the set x_k is called the Gauss grid. For the Chebyshev polynomial, the $Q(x)$ corresponding to Equation (4.6) can be given as

$$Q(x) = T_{N+1}(x) = 2^N \prod_{k=0}^{N} (x - x_k)$$ (4.16)

On the Gauss grid, it can be shown that the interpolation error bound (Geurts 2004) is given by

$$\|u - (I_N u)\| = \max_{-1 \leq x \leq 1} |u(x) - (I_N u)(x)| \leq \frac{1}{2^N (N+1)!} \|u^{(N+1)}\|$$ (4.17)

Following the derivations of Equations (4.12) and (4.13), for the Chebyshev polynomial, the elements of the first derivative matrix can be derived as

$$d_{ij} = \frac{(-1)^{i-j}}{x_i - x_j} \sqrt{\frac{1 - x_j^2}{1 - x_i^2}}; \quad i \neq j$$ (4.18)

$$d_{jj} = \frac{x_j}{2(1 - x_j^2)}$$ (4.19)

From Equations (4.18) and (4.19), the entries for the second derivative matrix can also be found. The disadvantage of the Gauss grid for the Chebyshev polynomial is that the boundaries themselves do not enter the grid. An alternative grid, referred to as the Gauss-Lobatto grid (Canuto et al. 1988), including the boundaries, is given by $x_k = \cos(k\pi/N)$ for $k = 0,1,...,N$ and is built on $Q(x) = (x^2 - 1)T_N'(x)$. Canuto et al. (1988) also discussed further extensions involving additional coordinate transformations to accommodate more complex and/or semi-infinite flow domains. The Chebyshev polynomials are quite different from those in other discretization methods such as finite difference schemes, where uniformity of the grid typically gives rise to an increased accuracy. For the approaches based on the Chebyshev polynomials, optimal performance of the numerical methods is achieved by using a nonuniform grid such as a Gauss grid or a Gauss–Lobatto grid.

Some advantages of a Chebyshev pseudospectral method include the very high accuracy associated with the absence of numerical dissipation and dispersion, high resolution near boundaries such as no-slip walls, and exponential convergence (Canuto et al. 1988). Unlike the pure spectral method using Fourier modes for periodic boundaries, Chebyshev polynomials can be applied for nonperiodic boundaries. However, such a method does not have flexibility in specifying the grid distribution according to the change of the flow field. Using Chebyshev polynomials, the specification of boundary conditions such as inflow/outflow boundaries is extremely difficult. The difficulty is associated with the choice of basis functions that satisfy the boundary conditions. Consequently, the Chebyshev polynomial expansion has been predominantly used in the derivative evaluations of the wall-normal direction in many DNS applications.

B. The Fractional Step Method

The issues of accuracy, stability, and computational speeds are relevant to both spatial discretization and time integration. Numerical methods for time integration are an integral part of DNS and LES, which have to be time-dependent simulations. Chapter 3 was devoted to time integration methods, particularly to high-order time integration methods suitable for DNS and LES. Most of the time integration methods used in the examples presented for DNS of incompressible channel flows were already covered in that chapter. An exception is the fractional step method for incompressible flows, which has been broadly used in the time-advancement of incompressible DNS (e.g., Redjem-Saad et al. 2007; Yamamoto et al. 2000), but not covered in Chapter 3. In the following, the fractional step method for time advancement is briefly presented.

For incompressible flows, the fundamental governing equations lack an independent equation for the pressure. The continuity equation cannot be used directly. The fractional step method solves the governing equations for incompressible flows in an effective manner, which forces the continuity equation to be satisfied. In this method, first an intermediate velocity field is found that does not normally satisfy continuity, and then by using this velocity field an equation for a virtual scalar quantity is found that is related to the pressure. From this scalar both the final velocity field and the pressure can be calculated, as described by Kim and Moin (1985).

The fractional step method is also referred to as the projection method, first proposed by Chorin (1968) and Temam (1969) and later successfully applied to the simulation of unsteady flow problems by Kim and Moin (1985). There are various versions of fractional step methods, slightly differing from each other. In the fractional step method, a common practice for the temporal discretization of the unsteady incompressible Navier–Stokes equations is to apply an explicit scheme for the convection terms and an implicit scheme for the diffusion terms. The explicit treatment of the nonlinear terms eliminates the need for linearization, whereas the implicit treatment of the diffusion terms eases the numerical stability restriction.

As demonstrated by Kim and Moin (1985), the second-order Adams–Bashforth scheme for the convection terms and the second-order Adams–Moulton scheme for the diffusion terms were used for the temporal discretization of the unsteady incompressible Navier–Stokes momentum equations. The fractional step method essentially decoupled the pressure and velocity, which obtained the time-dependent pressure and the divergence-free velocity satisfying the continuity equation. With this approach, the discretized governing equations can be expressed as

$$\frac{\tilde{u}_i^{n+1} - u_i^n}{\Delta t} = \frac{1}{2}\left(3C_i^n - C_i^{n-1}\right) + \frac{1}{2}\left(\tilde{D}_i^{n+1} + D_i^n\right) \tag{4.20}$$

$$\frac{u_i^{n+1} - \tilde{u}_i^{n+1}}{\Delta t} = -\frac{\partial \phi^{n+1}}{\partial x_i} \tag{4.21}$$

$$\frac{\partial u_i^{n+1}}{\partial x_i} = 0 \tag{4.22}$$

In Equations (4.20)–(4.22), $\tilde{u}_i$ is the intermediate velocity component and u_i^{n+1} is the final velocity field with superscript $n+1$ representing the current time step, ϕ is a scalar related to pressure, C_i represents the convection term, and D_i represents the diffusion term. The detailed forms of these terms can be found in Kim and Moin (1985), which can also be readily obtained from the fundamental governing equations for an incompressible flow presented in Chapter 1. Equations (4.21) and (4.22) can be combined to eliminate u_i^{n+1}, leading to the pressure Poisson equation given in terms of ϕ. Once this pressure Poisson equation is solved, u_i^{n+1} can be obtained from Equation (4.21).

Therefore, the time advancement of the fractional step consists of the following steps:

- Step 1. Calculate a predicted velocity or an intermediate velocity $\tilde{u}_i^{n+1}$ from Equation (4.20).

- Step 2. Solve the Poisson equation for the (modified) pressure.

- Step 3. Correct the velocity to enforce mass conservation using Equation (4.21).

Numerical solution of the Poisson equation is an integral and important part of the numerical procedure. There are a variety of methods that can be used to solve the pressure Poisson equation. The commonly used method is based on the traditional tridiagonal matrix algorithm (TDMA), also known as the Thomas algorithm, which is a simplified form of Gaussian elimination that can be used to solve tridiagonal systems of equations. The conventional TDMA or Thomas algorithm directly solves the discretized equations in one dimension and can be applied iteratively, in a line-by-line fashion, in the other two dimensions to solve multidimensional problems. When compared against direct solving methods, the conventional TDMA is computationally inexpensive and has the advantage of using the minimum amount of memory storage. So far, the conventional TDMA solver has been used for a wide variety of applications in spatial flow simulations in which the discretized equations can be reduced to a tridiagonal form in each of the three dimensions of space. For temporal DNS with periodic boundaries, spectral methods as discussed before can also be conveniently used to solve the pressure Poisson equation.

In the early version of the fractional step method given by Kim and Moin (1985) and the version of Xu et al. (2005), the pressure was neglected in the first step of the method and there might be some difficulties to assign the correct boundary conditions to the intermediate velocity field. To avoid this, the pressure at the previous time step may be used explicitly in the first step, which makes the boundary conditions of the intermediate velocity the same as the real velocity field. Finally, it is worth mentioning that the fractional step method of Kim and Moin (1985) is a second-order scheme. Higher-order fractional step methods can be achieved by involving higher-order time-integration schemes such as those discussed in Chapter 3.

REFERENCES

Canuto, C., Hussaini, M.Y, Quarteroni, A., and Zang, T.A. 1988. *Spectral methods in fluid dynamics.* New York: Springer-Verlag.

Chorin, A.J. 1968. The numerical solution of the Navier–Stokes equations. *Mathematics of Computation* 22: 745–762.

Ekaterinaris, J.A. 2005. High-order accurate, low numerical diffusion methods for aerodynamics. *Progress in Aerospace Sciences* 41: 192–300.

Geurts, B.J. 2004. *Elements of direct and large-eddy simulation.* Philadelphia PA: Edwards.

Iida, O., Kasagi, N., and Nagano, Y. 2002. Direct numerical simulation of turbulent channel flow under stable stratification. *International Journal of Heat and Mass Transfer* 45: 1693–1703.

Kim, J. and Moin, P. 1985. Application of a fractional-step method to incompressible Navier–Stokes equations. *Journal of Computational Physics* 59: 308–323.

Lele, S.K. 1992. Compact finite-difference schemes with spectral-like resolution. *Journal of Computational Physics* 103: 16–42.

Marquillie, M., Laval, J.-P., and Dolganov, R. 2008. Direct numerical simulation of a separated channel flow with a smooth profile. *Journal of Turbulence* 9(1): 1–13.

Redjem-Saad, L., Ould-Rouiss, M., and Lauriat, G. 2007. Direct numerical simulation of turbulent heat transfer in pipe flows: Effect of Prandtl number. *International Journal of Heat and Fluid Flow* 28: 847–861.

Temam, R. 1969. Sur l'approximation de la solution des equations de Navier–Stokes par la méthode des pas fractionnaires, i et ii. *Archive for Rational Mechanics and Analysis* 32: 135–153.

Xu, H., Yuan, W., and Khalid, M. 2005. Design of a high-performance unsteady Navier–Stokes solver using a flexible-cycle additive-correction multigrid technique. *Journal of Computational Physics* 209: 504–540.

Yamamoto, Y., Kunugi, T. and Serizawa, A. 2000. Direct numerical simulation of coherent structure in turbulent open-channel flows with heat transfer. In *Lecture Notes in Computer Science*, eds. Valero, M. et al., 502–513. Berlin, Germany: Springer-Verlag.

CHAPTER 5

DNS of Compressible Flows

IN THE PREVIOUS CHAPTER, sample DNS results of incompressible channel flows were presented, with discussions given on the numerical features of the methods used in the simulations. In contrast to incompressible flows, which are of relatively low speeds, compressible flows represent a broad range of fluid flows with relatively high speeds, ranging from low compressible subsonic flows with Mach numbers slightly higher than 0.3, to highly compressible hypersonic flows with Mach numbers exceeding 5. Compressible flows have many applications, especially in aerospace engineering. DNS provides a very powerful tool to investigate compressible flows and to gain insight into the fundamentals of the flow, turbulence, and mixing processes at high speeds. An enhanced understanding obtained by DNS can help many practical applications in different ways, including the development of effective flow control techniques for aeronautics and astronautics. DNS results can be effectively utilized to interpret available experimental data, to guide experimental work, and to execute calculations for operating conditions that are not achievable using experimental methods. In many applications of compressible flows, DNS can provide results that are not possible or are difficult to obtain using experimental and/or analytical methods and other computational methods.

Compressible flows are generally high-speed gas flows and the pivotal aspect of high-speed flows is that the density is variable. In addition to variable density, another pivotal aspect of high-speed compressible flow is energy. The energy of a given molecule is the sum of its transitional, rotational, vibrational, and electronic energies. For compressible flows, the most useful quantity to represent energy is the internal energy, which is defined as the sum of the energies of all the molecules in a finite volume of gas consisting of a large number of molecules. The internal energy per unit mass of gas is defined as the specific internal energy, denoted by e. A high-speed flow is a high-energy flow. When the flow velocity is increased, some of the kinetic energy is lost and reappears as an increase in internal energy, hence increasing the temperature of the gas. Therefore, in a high-speed flow, energy transformations and temperature changes are important considerations, whether the flow is reacting with combustion heat release or nonreacting without combustion heat release. In contrast to compressible flows, the energy equation for incompressible flows is important only to heat transfer problems or reacting flows such as combustion applications.

The considerations of energy for compressible flows come under the science of thermodynamics. In thermodynamics, another variable used to represent energy is enthalpy h, apart from internal energy e, which can be defined as $h = e + pv$, where p is thermal dynamic pressure, and v is the specific volume or the volume per unit mass. For compressible flows, the thermodynamic properties of the gas are of great importance where the fluid can be very often assumed as a perfect gas—a gas in which the intermolecular forces are neglected. For a perfect gas, the equation of state is $p = \rho RT$, or $pv = RT$ (where R is the gas constant and T is temperature). For a perfect gas, it can be readily shown that $c_p - c_v = R$, $c_p = \gamma R/(\gamma - 1)$ and $c_v = R/(\gamma - 1)$, where $\gamma = c_p/c_v$ is the ratio of specific heats at constant pressure and volume. For a perfect gas, e and h are functions of temperature only: $de = c_v\, dT$, $dh = c_p\, dT$. A perfect gas where c_v and c_p are constants is defined as a calorically perfect gas: $e = c_v T$, $h = c_p T$. Note that e and h are thermodynamic state variables and have nothing to do with the fluid dynamic process that may be taking place. These thermodynamic equations are frequently involved in the numerical solution of a compressible flow field, as auxiliary equations to the fundamental fluid dynamic governing equations given in Chapter 1.

The most convenient index to gage whether a fluid flow must be considered as compressible is the Mach number, defined as the ratio of the local velocity to the local speed of sound. The physical mechanism of sound propagation in a gas medium is based on molecular motion (by molecular collisions). Common experience tells us that sound travels through air at some finite velocity. The speed of sound in a calorically perfect gas (a perfect gas where c_v and c_p are constants) is given by $a = \sqrt{\gamma p/p} = \sqrt{\gamma RT}$. Mach number can also be used to define different regimes of compressible flows: (1) subsonic flow: $M_\infty < 0.8$; (2) transonic flow with

$M_\infty < 1 :$ $0.8 < M_\infty < 1$; (3) transonic flow with $M_\infty > 1$: $1 < M_\infty < 1.2$; (4) supersonic flow: $M_\infty > 1.2$; and (5) hypersonic flow: $M_\infty > 5$. For practical applications, a compressible flow field normally covers several regimes with different speeds. As the flow changes from subsonic to supersonic, the complete nature of the flow changes, not the least of which is the occurrence of shock waves. The formation of shock waves is associated with two physical conditions: a supersonic freestream velocity and the presence of an obstacle in the flow field such as a solid body. In practical applications, there is always the presence of solid bodies in the flow field such as wings or fuselage of the aircraft, vessels in aircraft engines, diffusers and nozzles, and the ground effect. In a fluid flow environment, the disturbances generated due to the presence of the solid body are propagated upstream via molecular collisions at approximately the speed of sound. At supersonic speeds, the flow moves faster than the speed of sound; therefore, disturbances cannot eventually work their way upstream. Instead, they coalesce, forming a standing wave that is the shock wave. A shock wave is an extremely thin region, typically on the order of 10^{-5} cm, across which the flow properties can change drastically. The flow across a shock wave may be regarded as adiabatic (no heating or cooling). Across a shock wave in the streamwise direction, the pressure, density, temperature, and entropy increase; the Mach number, flow velocity, and total pressure decrease; and only the total enthalpy stays the same. Physically, a shock wave is a very thin region across which some large changes in fluid properties occur almost discontinuously, and within the shock wave itself, large gradients in velocity and temperature occur; that is, the mechanisms of friction and thermal conduction are strong. These are dissipative, irreversible mechanisms that always increase the entropy. Unlike the sound wave, the flow through the shock wave is nonisentropic. The shock wave is referred to as a normal shock wave if it stands in a direction normal to the streamwise direction. In general, a shock wave makes an oblique angle with respect to the upstream flow since the surface of the solid body does not always present in a direction normal to the streamwise velocity. In such an oblique shock wave, the pressure also increases discontinuously across the wave. The oblique shock waves are inherently multidimensional in nature, involving sharp gradients. Apart from shock waves, supersonic flows are also characterized by expansion waves, where the pressure decreases continuously across the wave. Oblique shock waves occur when a supersonic flow is turned into itself. In contrast, when a supersonic flow is turned away from itself, an expansion wave is formed. For supersonic compressible flows, both external flows such as the flow around a finite wing or a jet flow in an open boundary domain and internal flows such as the flow through nozzles, diffusers, and wind tunnels can develop complex shock and expansion wave structures.

DNS of compressible supersonic flows involving shock waves are very challenging for numerical simulations because the numerical methods have to fully resolve the sharp gradients in the flow regions involving shock waves, which is very often impossible due to the prohibitive computational costs. In the general CFD area, many efforts have been devoted to developing approximate methods for flow regions involving shock waves in order to obtain a reasonable representation of the overall flow field rather than a detailed solution of the shock waves. In fact, the detailed shock wave structures are very often neglected. Many computational methods have been developed for the solution of compressible flows involving shock waves, such as the shock-fitting method and shock-capturing methods. In these efforts, the governing equations solved are the inviscid governing equations—the hyperbolic Euler equation described in Chapter 1, where the viscous effects are neglected. In general CFD for compressible flows using the shock-fitting method, shock waves are explicitly introduced in the solution using appropriate shock relations such as the Rankine-Hugoniot relations. The shock-fitting method is an elaborate method based on the analytical solutions of shock waves, which is in contrast to the shock-capturing method. The shock-capturing approach represents a class of techniques for computing inviscid flows with shock waves in which the governing equations of inviscid flows are cast into the conservation form and any shock waves or discontinuities are computed as part of the solution. In such an approach, no special treatment is employed to take care of the shocks themselves. In concept, the shock-capturing methods are relatively simple compared to the more elaborate shock-fitting methods. However, the shock waves predicted by shock-capturing methods are generally not sharp and smear over several grid points. Also, classical shock-capturing methods have the disadvantage that unphysical oscillations may develop in the vicinity of strong shocks. In traditional CFD for compressible flows, symmetric or central schemes that are normally of second-order accuracy are used and do not consider any information about wave propagation in the discretization. In order to avoid numerical oscillations in the shock regions, artificial viscosity may be added onto the discretized Euler equations. In the classical shock-capturing methods, numerical dissipation terms are usually linear and the same amount is uniformly applied at all grid points. These classical shock-capturing methods exhibit accurate results only in the case of smooth and weak-shock solution; but when strong shock waves are present in the solution, nonlinear instabilities and oscillations can arise across discontinuities. Apparently, these methods are not robust enough. In contrast to these classical methods, modern shock-capturing methods are generally upwind based, where the differencing schemes attempt to discretize the hyperbolic partial differential equations by using differencing biased in the direction determined by the sign of the characteristic speeds. No matter what type of shock-capturing scheme is used, a stable calculation for flows with the presence of shock waves requires a certain amount of numerical dissipation in order to avoid the formation of unphysical numerical oscillations. In modern shock-capturing methods, a nonlinear numerical dissipation is added, with an automatic feedback mechanism that adjusts the amount of dissipation in any cell or at any grid point of the mesh, in accord with the gradients in the solution. These schemes have proven to be stable and accurate even for problems containing strong shock waves. Some of the well-known classical shock-capturing methods include the MacCormack method, Lax–Wendroff method, and Beam–Warming method, as described by Anderson (2004) and Hirsch (2007). Examples of modern shock-capturing schemes include the widely used total variation diminishing (TVD) schemes first proposed by Harten (1983), the flux-corrected transport scheme introduced by Boris and Book (1976), the monotonic upstream-centered schemes for conservation laws (MUSCL) based on the Godunov approach and introduced by van Leer (1979), the approximate Riemann solvers proposed by Roe (1981), the various essentially nonoscillatory (ENO) schemes proposed by Harten et al. (1987) and discussed also by Shu and Osher (1988), and the piecewise parabolic method (PPM) proposed by Colella and Woodward (1984). These shock-capturing methods have been proved effective, but the numerical accuracy at locations near the shock waves inevitably decays.

As far as DNS of compressible flows is concerned, strictly speaking, the inviscid Euler equations are not so relevant, due to the fact that these equations neglect the dissipative effects, which are the most important characteristic of small-scale turbulence. The Kolmogorov microscales, which in principle should be resolved in DNS, are decided by the viscous dissipation effects. In DNS, the viscous effects should always be considered. In traditional CFD for high-speed compressible flows, the simulations were predominately carried out for inviscid flows, which cannot be regarded as DNS although high-order numerical schemes might have been used. To date, the applications of DNS to compressible flows have been mainly restricted to subsonic flows with relatively low speeds, due to the prohibitive costs of the computations at higher flow speeds. This chapter is restricted to simulations of subsonic flows, where the formation of shock waves is not prevalent. The governing equations used are the compressible Navier–Stokes equations with the viscous terms, instead of the hyperbolic systems that are frequently employed in traditional CFD for high-speed compressible flows.

In the context of DNS of subsonic compressible flows, one example is DNS of jet flows, which are of great importance to many practical applications such as propulsion and noise generation. As a class of free shear layer flow, jets are also of great importance to fundamental fluid dynamics research. Over the last two decades, there has been a large amount of DNS of compressible jet flows. This chapter presents a few applications of DNS to compressible jet flows with some sample results of DNS of both nonreacting and reacting jet flows. Simulation results for both round jets and noncircular jets are presented. Techniques for DNS data postprocessing such as the proper orthogonal decomposition (POD) techniques are also discussed. Finally, discussions on the numerical features of the simulations are presented.

I. SAMPLE RESULTS: DNS OF COMPRESSIBLE JET FLOWS

A. DNS of a Compressible Plane Jet (Reichert and Biringen 2007)

Compressible jet flows such as flows established on nozzles have important engineering applications. The compressible plane jet issuing into a parallel stream is a flow analogous to exhaust from a rectangular nozzle of high aspect ratio of an aircraft engine in forward motion. The study of Reichert and Biringen (2007) involved an unconfined plane jet in a coflowing stream with hyperbolic tangent shear layer profiles at the inflow. In their simulations, the instability was triggered using random perturbations at the inflow. A spatial compact finite difference scheme with Runge–Kutta time advancement was utilized for the integration of the governing equations. The nonreflecting characteristic boundary conditions were employed at the side boundaries. Figure 5.1 shows the computational geometry and coordinate system. Both two-dimensional and fully three-dimensional simulations have been performed.

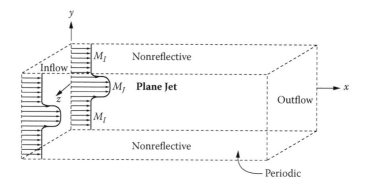

FIGURE 5.1 Computational geometry and coordinate system: DNS of a compressible plane jet. (Reichert and Biringen 2007; with permission from Elsevier Science Ltd.)

Figure 5.2 shows the instantaneous spanwise vorticity contours from the two-dimensional simulations at various Mach numbers. The first notable feature is the large degree of organization visible in the far downstream region. Large-scale vortical structures develop, while an increase in Mach number lengthens the jet potential core with the shear layers becoming

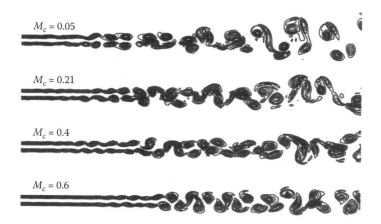

FIGURE 5.2 Instantaneous spanwise vorticity contours at different Mach number from the two-dimensional simulations: DNS of a compressible plane jet. (Reichert and Biringen 2007; with permission from Elsevier Science Ltd.)

(a) *Ma* = 0.4

(b) *Ma* = 0.7

FIGURE 5.3   Instantaneous vorticity magnitude visualizations in the *xy*-plane from the three-dimensional simulations: DNS of a compressible plane jet. (Reichert and Biringen 2007; with permission from Elsevier Science Ltd.)

wavy. Roll-up and pairing of the vortical structures occur further downstream. The downstream cross-streamwise spreading of the jet is reduced with increasing Mach number. The highest Mach number case exhibits more stable vortex streets leading to less pairing and decreased spreading, suggesting less mixing with increased compressibility. In order to examine the three-dimensional effect, Figure 5.3 shows the instantaneous vorticity distributions from the three-dimensional simulations for *Ma* = 0.4 and *Ma* = 0.7. The instantaneous vorticity visualizations of Figure 5.3, with lighter regions indicating higher vorticity, reveal several key points. In the upstream region, the *xy*-cuts in Figure 5.3 resemble the two-dimensional vorticity plots of Figure 5.2. The shear layers become wavy and then roll up into discrete vortices. The antisymmetric mode clearly dominates the initial vortex street, as evidenced by the spatially staggered trains of vortex cores. In addition, there is some indication of vortex pairing and the associated increase in length scales. However, the similarities end in the downstream region. Whereas the two-dimensional simulations predict multiple downstream amalgamations into strong vortices, the three-dimensional calculations display a rapid breakdown to smaller, less organized motions.

It is worth noting that in the simulations performed by Reichert and Biringen (2007), the Euler equations, which express conservation of mass, momentum, and energy for the inviscid motion of compressible fluids, were employed. As mentioned before, the simulations may not be regarded as "DNS" if viscous effects were ignored since the Kolmogorov microscales are defined based upon viscous effects. In principle, the grid resolution used in DNS should be smaller than the smallest scale in the flow field, the Kolmogorov length scale, for it to be resolved. To capture the smallest scales in the flow field is a very stringent requirement for a numerical simulation, and is always difficult to satisfy due to the prohibitively large number of grid points needed. Nevertheless, the terminology of "DNS" has been broadly used (or misused) in modern CFD literature without strictly following the definition of resolving all the relevant time and length scales. This may not satisfy a purist; however, one could argue that this terminology is not clear-cut, since very often the smallest scale, or the Kolmogorov microscale, is not a known quantity for a practical simulation. Of course it may be known from experiments or another simulation, or it may be estimated based on the amount of fluctuation observed in the simulation results. It would be more appropriate to examine the numerical credibility of the simulation rather than the accuracy of the usage of the terminology. In addition, two-dimensional simulations were performed by Reichert and Biringen (2007), which ideally should not be regarded as "DNS" due the lack of three-dimensional vortex stretching and interaction. Physically, there is no such thing as two-dimensional turbulence, albeit with the fact that there are many simulations performed in the reduced dimensions to reduce the computational costs. From a personal point of view, the authors tend to accept that a simulation may be regarded as DNS-like if the following conditions are satisfied: (1) the time-dependent, three-dimensional governing equations based on mass, momentum, and energy conservations are solved without modeling (or approximation) the fluid motion, or, in other words, a turbulence model is not used; instead, any turbulence present in the flow field should be a natural consequence of the numerical solution; and (2) the numerical solutions should be sufficiently accurate without appreciable errors from the numerical solvers, including the numerical schemes and boundary conditions used. The numerical schemes should be free of dissipation and dispersion errors, or at least have only a minimal amount of these errors. To achieve this, highly accurate numerical methods such as spectral methods or high-order schemes should always be used. Furthermore, boundary conditions need to be able to represent the physical conditions as faithfully as possible, while not introducing any numerical inaccuracy into the numerical solution inside the computational domain.

B. DNS of a Variable-Density Round Jet (Nichols et al. 2007)

Fully three-dimensional DNS of variable-density round jets with and without gravity was performed by Nichols et al. (2007). The compressible time-dependent Navier–Stokes equations were solved in cylindrical coordinates on a staggered grid. Sixth-order compact finite difference schemes were used to compute the spatial derivatives. A third-order Adams–Bashforth time-differencing scheme was used to advance the advection and diffusion terms in the momentum and energy equations. An asymptotic method was employed to treat the centerline singularities at $r = 0$. Viscous, traction-free, open-boundary conditions were applied at the cross-streamwise and outflow boundaries. At the outflow, a free-slip collar was also used, which is very similar to the sponge layer used by other researchers such as Jiang and Luo (2000), to control the wave reflections from outside the computational domain.

Figures 5.4 and 5.5 show the time development of the jet from initial conditions for two different Froude numbers (dimensionless parameters comparing inertial and gravitational forces): $Fr = \infty$ (the nonbuoyant case) and $Fr = 8.0$ (the buoyant case). Each successive image is separated from the previous one by a nondimensional time of 12. In both cases, the flow initially adjusts to become nonparallel. By nondimensional time $t = 48$ shown in Figures 5.4(e) and 5.5(e), however, the initial adjustment has left the computational domain. In the final images of both sequences, it can be observed that the laminar inflow at the bottom of the domain enters first into an oscillatory state that is dominated by a train of Kelvin–Helmholtz type vortices convecting downstream along the shear layer. In the case of the nonbuoyant jet, long, axially aligned streak-like structures are observed to develop in Figures 5.4(e)–(g). These structures are sometimes referred to as side jets because they form by jet fluid being ejected from the jet center in a star-shaped pattern. Side-jet formation is markedly absent in the case of the buoyant jet, the results of which show that the transition to the quasi-turbulent flow observed at the top of the domain is much more abrupt in the case of the buoyant jet.

C. DNS of a Transitional Rectangular Jet (Rembold et al. 2002)

Jets established on noncircular nozzles such as rectangular jets provide an effective passive flow control means in many applications due to the enhanced mixing in comparison with round jets. DNS of a Mach 0.5 jet exiting from a rectangular-shaped nozzle with an aspect ratio of 5 into

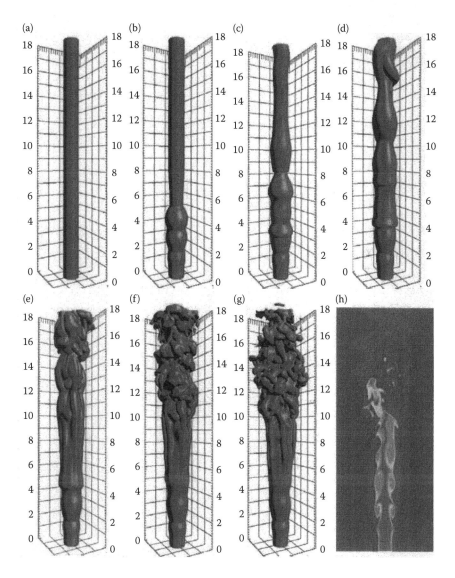

FIGURE 5.4   Time series of the jet evolution without gravity $(Fr = \infty)$: DNS of a variable-density round jet. (Nichols et al. 2007; with permission from Cambridge University Press.)

a quiescent ambient was performed by Rembold et al. (2002). The three-dimensional compressible Navier–Stokes equations were solved on a Cartesian grid. Spatial discretization was achieved by using a fifth-order compact upwind-biased scheme for the convective terms and a sixth-order compact central scheme for the diffusive terms. Time-integration

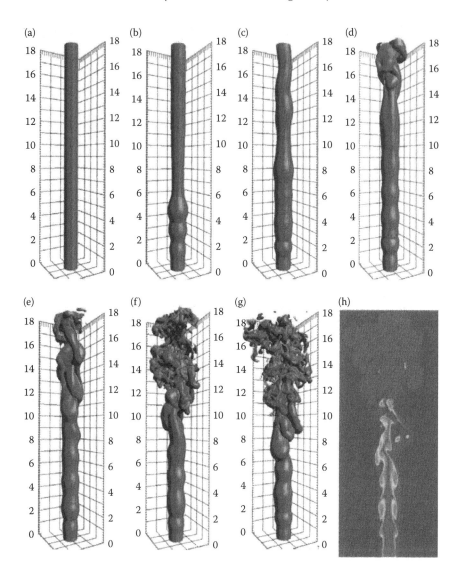

FIGURE 5.5 Time series of the jet evolution without gravity ($Fr = 8.0$): DNS of a variable-density round jet. (Nichols et al. 2007; with permission from Cambridge University Press.)

was performed with a third-order low-storage Runge–Kutta scheme. The nonreflecting characteristic boundary conditions were applied at the side boundaries. A sponge layer was employed at both the inflow and outflow boundaries to prevent spurious wave reflections. At the inlet, Dirichlet conditions were used.

FIGURE 5.6 Snapshot of density isosurface: DNS of a transitional rectangular jet. (Rembold et al. 2002; with permission from Elsevier Science Ltd.)

In Figure 5.6, the instantaneous flow topology of the transition process is visualized by a snapshot of the density isosurface. A strong growth of the linear instability mode in the laminar region is observed. Rapidly three-dimensional disturbances of the initially two-dimensional instability mode grow at the lateral edges of the jet. A symmetric vortex structure develops but breaks up almost immediately into small-scale turbulence. Figure 5.7 shows a close-up of a pressure isosurface in the transition region. Vortices symmetric to the x-y plane can be identified that deform and finally break up into disordered structures. Additionally, Figure 5.8

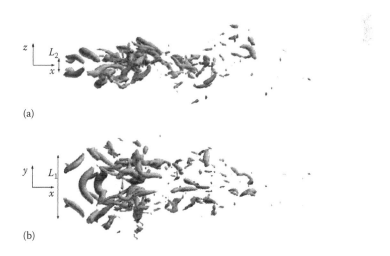

FIGURE 5.7 Snapshot of pressure isosurface in the transition region. (a) Side view and (b) top view, flow from left to right: DNS of a transitional rectangular jet. (Rembold et al. 2002; with permission from Elsevier Science Ltd.)

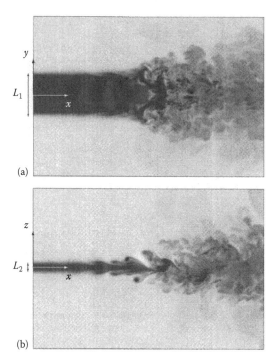

(a)

(b)

FIGURE 5.8 Snapshot of density contours along the (a) major and (b) minor axis of the jet at the same instant: DNS of a transitional rectangular jet. (Rembold et al. 2002; with permission from Elsevier Science Ltd.)

displays density plots in two cross-sections through the jet center. The flow shows a tendency of "axis-switching," which is a typical fluid dynamic behavior of rectangular jets. The excited varicose mode develops toward the antisymmetric sinuous mode, which is followed by a rapid transition to turbulence. The observed flow structure and transition location are the result of the imposed inflow disturbances and the spatial development of the flow instabilities.

D. DNS of a Swirling Annular Jet Flame

As an example of applications of DNS to compressible reacting flows, some sample results of a swirling annular nonpremixed flame obtained from the author's group at Brunel University are presented here. Swirling flows are encountered in applications such as engines, turbine combustors, furnaces, and boilers. Swirling motion is regarded as an efficient way to stabilize nonpremixed flames and has been used together with bluff-body stabilization. In such a configuration, a large vortex with flow reversal may be formed in the core region, which carries the combustion products back toward the burner mouth, providing a continuous and stable source of energy for flame ignition. In addition, swirl extends the curved shear layer and produces extra shear promoting turbulence generation that enhances mixing and combustion intensity. In many practical applications, it is preferable to have relatively strong swirl so that there is a formation of pronounced coherent structures involving both axisymmetric and azimuthal vorticity. In combustion applications, coherent vortical structures play an important role as they influence mixing of heat and species to a large extent and hence the entire reaction process. It is difficult to study swirling flows using the traditional Reynolds-averaged methods due to the existence of unsteady coherent structures and the effects of mean flow streamline curvature (Jakirlić et al. 2002). Recently there has been a substantial amount of LES on swirling flames (e.g., Duwig and Fuchs 2005; Grinstein and Fureby 2005; Sankaran and Menon 2002; Selle et al. 2006; and Stein et al. 2007; to name but a few). In most of these studies, the swirling flame was established on a round nozzle. For flows established on a round nozzle, there is a circular velocity shear layer near the nozzle exit where the flow instability may develop into vortical structures and turbulence downstream. Different from a round nozzle, an annular nozzle has two adjacent velocity shear layers, which may enhance the turbulence levels within a flame and improve mixing and combustion and reduce pollutant emissions.

Flames in annular configurations are used in industrial and domestic burners such as cooking flames. For an annular jet, the fluid dynamics is largely determined by the interaction of two adjacent circular shear layers (Patte-Rouland et al. 2001), which is significantly different from that of a round jet with one such shear layer. The prominent features of annular jet flames include the formation of a recirculation zone inside the jet core near the nozzle exit, which not only enhances the fuel/air mixing, but also stabilizes the flame. Vanoverberghe et al. (2003) experimentally investigated a swirl-stabilized partially premixed combustion in an annular configuration in a confined environment, where different flame states were observed. On the computational side, large-eddy simulation has been used to simulate annular nonreacting swirling jets (e.g., García-Villalba and Fröhlich 2006; García-Villalba et al. 2006). The LES captured the precessing vortex cores

(PVC), which are the inner structures in swirling jets, featuring the vortex core rotating and vortex spinning. However, LES or DNS of reacting flows such as flames in annular configurations has not been available.

Another important topic in combustion research is near-wall combustion, which deserves more research efforts from both application and fundamental points of view. In this context, the study of impinging jet flames is of particular interest. In addition to the relevance to many engineering applications such as industrial burners, metal cutting, glass shaping, and glass melting for fiber optics production, impinging jet flames are also of great value in fundamental academic studies. The impinging flow configuration is of simple geometry but covers a broad range of important flow phenomena, such as large- and small-scale structures, wall boundary layers with stagnation, large curvature involving strong shear and normal stresses, and wall heat transfer. Impinging flames involve the complex interactions between the wall and the flame. In a turbulent scenario, the flame/wall/turbulence multiway interactions bring many unresolved and challenging issues to combustion modeling (Poinsot and Veynante 2001). In general, near-wall flow and heat transfer in reacting flows are not well investigated. For instance, the classical law-of-the-wall models of fluid flow and heat transfer neglected the presence of flame and variable density effects, which should be taken into account for combustion applications (Jiang et al. 2007). Due to the rich flow phenomena involved and the geometric simplicity, the impinging flame is ideal for development and validation of near-wall models. The impinging flow contains a broad range of length scales, ranging from the large-scale vortical structure to very thin thermal boundary layers near the wall. The near-wall flow and combustion processes are also highly unsteady. A detailed study of impinging flames requires both temporally and spatially resolved solutions, and DNS provides such a possibility.

In the study carried out at Brunel University, the dynamics of annular swirling nonpremixed jet flames has been investigated by performing 3D parallel DNS. Highly accurate numerical methods and high-fidelity boundary conditions have been used in the DNS. The swirling motion was introduced into the annular fuel jet itself. In such a configuration, the burner surface area inside the jet annulus provides bluff-body stabilization for the nonpremixed flame. In order to examine the swirling effects on the flow and flame dynamics, two cases have been performed, including one case without swirl and another case with swirl number 0.4. In both cases, a wall at the ambient temperature is present at the downstream location. The physical problem considered was a fuel jet from an annular nozzle issuing into an open boundary domain with downstream wall confinement and a swirl applied onto the fuel stream. Combustion took place when the fuel mixed with the oxidant environment. The computational domain was the region between the jet nozzle exit plane and the downstream impinging wall, which is open to the ambient environment in the cross-streamwise directions.

The mathematical formulation included the governing equations, numerical methods for discretization, and solution and boundary conditions. The flow was described with the compressible time-dependent Navier–Stokes equations in the Cartesian coordinate system (x, y, z), where the z axis is along the streamwise direction of the fuel jet and the $x - 0 - y$ plane is the domain inlet where the jet nozzle exit locates. The nondimensional form of the governing equations was employed (Jiang and Luo 2003). Major reference quantities used in the normalization were the maximum streamwise mean velocity at the jet nozzle exit (domain inlet), nozzle diameter (measured from the middle points of the annulus), and the ambient temperature, density, and viscosity. Since this study was focused on the investigation of the fluid dynamic behavior of the flame, a simple chemistry $v_f M_f + v_o M_o \rightarrow v_p M_p$ with finite-rate Arrhenius kinetics was considered to be adequate, where M_i and v_i represent the chemical symbol and stoichiometric coefficient for species i, respectively. The reaction rate, after normalization, takes the form of $\omega_T = Da(\rho Y_f/W_f)^{v_f}(\rho Y_o/W_o)^{v_o} \exp[-Ze(1/T - 1/T_{fl})]$, where W and Y represent species molecular weight and mass fraction, and Da, Ze, and T_{fl} stand for the Damköhler number, Zeldovich number, and flame temperature, respectively. The heat release rate in the energy equation was given by $\omega_h = Q_h \omega_T$, with Q_h representing heat of combustion. The governing equations were supplemented by the ideal-gas law for the mixture. The equations were solved using a sixth-order accurate compact finite difference scheme for evaluation of the spatial derivatives (Lele 1992). The finite difference scheme allows flexibility in the specification of boundary conditions for minimal loss of accuracy relative to spectral methods. The time-dependent governing equations were integrated forward in time using a fully explicit low-storage third-order Runge–Kutta scheme.

Boundary conditions for the 3D spatial DNS of annular jet flames represent a challenging problem. Physical conditions at the nozzle exit must be appropriately represented. In the meantime, open boundary conditions in the jet cross-streamwise direction should allow jet mixing with the ambient and entrainment. The Navier–Stokes characteristic boundary condition (NSCBC) by Poinsot and Lele (1992) was utilized. For the nozzle exit (domain inlet), the NSCBC was used to specify the inflow boundary with density treated as a "soft" variable that fluctuated slightly in the simulations according to the characteristic waves at the boundary. To specify accurately the swirl number in a numerical simulation is a delicate issue. Based on the assumption of the equilibrium swirling inflow, analytical solution was derived for the mean velocity profiles (Jiang et al. 2008), including the mean streamwise velocity $\bar{w}$ and the mean azimuthal velocity $\bar{u}_\theta$:

$$\bar{w} = -\frac{1}{4}\frac{f_x}{\mu}\left(r^2 - \frac{R_i^2 - R_o^2}{\ln R_i - \ln R_o}\ln r + \frac{R_i^2\ln R_o - R_o^2\ln R_i}{\ln R_i - \ln R_o}\right) \tag{5.1}$$

$$\bar{u}_\theta = -\frac{1}{3}\frac{f_\theta}{\mu}\left(r^2 - \frac{R_i^2 + R_iR_o + R_o^2}{R_i + R_o}r + \frac{R_i^2 R_o^2}{R_i + R_o}\frac{1}{r}\right) \tag{5.2}$$

In Equations (5.1)–(5.2), $r = \sqrt{x^2 + y^2}$ is the radial distance, and R_i and R_o are the inner and outer radii of the annular jet, respectively. In Equations (5.1)–(5.2), f_x and f_θ can be defined by the maximum velocities at the inflow boundary. For a unit maximum velocity, which is often the case when a nondimensional form of the governing equations is employed, the constant f_x is defined as

$$f_x = -\frac{8\mu(\ln R_o - \ln R_i)}{R_o^2 - R_i^2 + R_i^2\ln\left[\dfrac{R_i^2 - R_o^2}{2(\ln R_i - \ln R_o)}\right] - R_o^2\ln\left[\dfrac{R_i^2 - R_o^2}{2(\ln R_i - \ln R_o)}\right] - 2R_i^2\ln R_o + 2R_o^2\ln R_i} \tag{5.3}$$

The parameter f_θ defines the degree of swirl. For known $\bar{w}$ and $\bar{u}_\theta$, the swirl number can be calculated from

$$S = \frac{\displaystyle\int_{R_i}^{R_o} \bar{w}\bar{u}_\theta r^2\,dr}{R_o\displaystyle\int_{R_i}^{R_o} \bar{w}^2 r\,dr} \tag{5.4}$$

A certain swirl number can be conveniently achieved by adjusting the constant f_θ in Equation (5.2) for $\bar{u}_\theta$. From the azimuthal velocity $\bar{u}_\theta$, the cross-streamwise velocity components at the inflow can be specified by $\bar{u} = -\bar{u}_\theta \, y/r$ and $\bar{v} = \bar{u}_\theta \, x/r$. The inlet velocity profiles have been perturbed by two small helical disturbances (Uchiyama 2004). The velocity components at the nozzle exit $z = 0$ were given as

$$u = \bar{u} + A \sum_m \sin(m\varphi - 2\pi f_0 t) \tag{5.5}$$

$$v = \bar{v} + A \sum_m \sin(m\varphi - 2\pi f_0 t) \tag{5.6}$$

$$w = \bar{w} + A \sum_m \sin(m\varphi - 2\pi f_0 t) \tag{5.7}$$

In Equations (5.5)–(5.7), A stands for the amplitude of disturbance, which was specified as 1% of the maximum value of streamwise mean velocity $\bar{w}$, m stands for the mode number, and φ stands for the azimuthal angle. In the simulations performed, two helical modes of $m = 1$ and $m = -1$ were superimposed on the temporal disturbance. The nondimensional frequency of the unsteady excitation was $f_0 = 0.30$, which was chosen to be the unstable mode leading to the jet preferred mode of instability. The fuel temperature at the inlet was chosen to ensure auto-ignition of the mixture.

Numerical results have been obtained for two computational cases with and without swirling. For the swirling case, the velocity of the fuel jet at the nozzle exit was specified with a swirl number of 0.4. In both cases, a downstream wall confinement was introduced into the computational domain. The considered jet Mach number was $M = 0.4$, and the Reynolds and Prandtl numbers used were Re = 2500 and Pr = 1. The ratio of specific heats used was $\gamma = 1.4$. The dynamic viscosity was chosen to be temperature dependent according to $\mu = \mu_a (T/T_a)^{0.76}$. Parameters used in the one-step chemistry were Damköhler number $Da = 6$, Zeldovich number $Ze = 12$, flame temperature $T_{fl} = 6$, and heat of combustion $Q_h = 150$. These values were chosen to give temperatures of a reacting flow typically encountered in many nonpremixed flames. The dimensions of the computational box used were $L_x = L_y = L_z = 6.0$. A grid system with 256 × 256 × 256 nodes was used with uniform distribution. A grid independence test was performed and further refinement of the grid by doubling the points

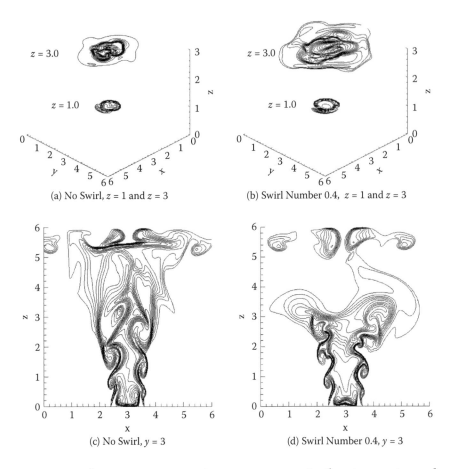

(a) No Swirl, $z = 1$ and $z = 3$

(b) Swirl Number 0.4, $z = 1$ and $z = 3$

(c) No Swirl, $y = 3$

(d) Swirl Number 0.4, $y = 3$

FIGURE 5.9 Instantaneous reaction rate contours in the streamwise and cross-streamwise sections at $t = 30$.

in each direction did not lead to appreciable changes in the solution. The time step was limited by the Courant–Friedrichs–Lewy condition for stability and a chemical restraint (Jiang and Luo 2003). The results obtained were considered to be grid and time-step independent. In the following, instantaneous results of the flow and combustion fields are presented and discussed first, followed by a proper orthogonal decomposition (POD) analysis of the instantaneous velocity field, and finally time-averaged results are presented and discussed.

Figure 5.9 shows the instantaneous reaction rate contours in the streamwise and cross-streamwise sections at $t = 30$ after the flame has been developed, for the cases without swirl and with swirl number 0.4. The cross-sectional locations of $y = 3$ and $z = 3$ are in the middle of the 3D domain. The results indicate that the annular nonpremixed flame develops complex dynamical structures. It is noticed that there is a flame located very close to the burner surface, inside the fuel annulus. This flame near the burner mouth can stabilize the combustion by providing a continuous and stable source of energy for ignition. The annular reacting flow also developed unsteady vortical structures, which are evident in the wrinkled flame structures in the jet columns. The vortical structures are caused by the Kelvin–Helmholtz type shear layer instability, which can play a significant role in the mixing and entrainment. In Figure 5.9, it is observed that, under the swirling condition, the flame becomes shorter but with larger spreading in the middle of the domain. Swirl extends the curved shear layer and promotes mixing and combustion; therefore, the flame becomes shorter and spreads more. In both the swirling and nonswirling cases, the flame touches the downstream wall boundary and quenches at the wall surface.

The annular jet flames develop into complex flow fields, as shown in the instantaneous velocity vectors in Figure 5.10. Under swirling conditions, the formation of the swirl-induced PVC can alter the flow pattern of the jet flame. For the nonswirling case, the flow field has a complex structure due to the development of the shear layer instability triggered by the helical modes in the small external perturbation applied at the nozzle exit. For the swirling case, an anti-clockwise flow motion can be observed near the center of the jet where the PVC locates. This anti-clockwise flow motion is caused by the external swirl at the nozzle exit. Around the rotating core, there are a few relatively weak structures associated with the helical modes. In Figure 5.10(d), it can be noticed that the center of the anti-clockwise flow motion is not exactly located at the geometrical center $(x = 3, y = 3)$ in the plane of $z = 3$, but it is roughly located at the geometrical center in the upstream plane of $z = 1$. The flow field of the swirling case indicates the existence of PVC, featuring the vortex core rotating associated with swirl. In both the swirling and nonswirling cases, at the downstream wall location of $z = 6$, the flow touches the impinging wall with the formation of a stagnation zone in the impinging region and wall jets in the surrounding areas, which are more evident in the nonswirling case.

In order to further examine the effect of the interaction between the unsteady vortical structures and the PVC associated with swirl, numerical

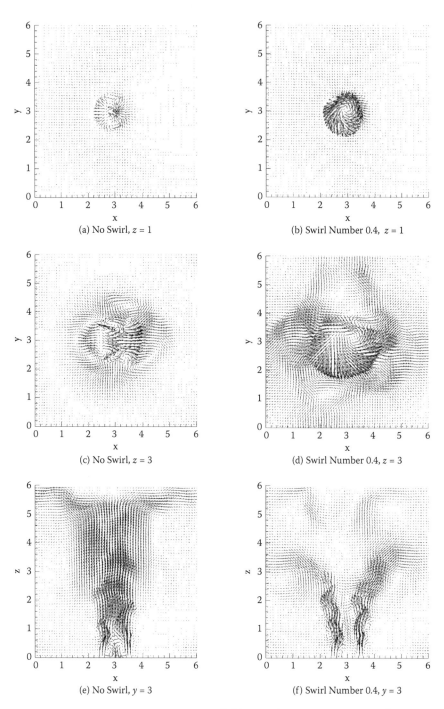

FIGURE 5.10 Instantaneous vector fields in the streamwise and cross-streamwise sections at $t = 30$.

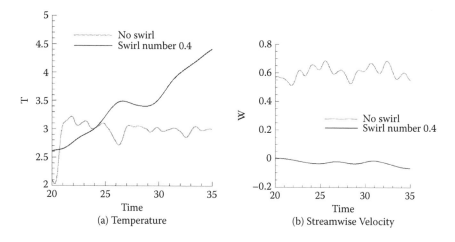

FIGURE 5.11   Time traces of the temperature and streamwise velocity at ($x =$ 3, $y = 3$, $z = 3$).

tests were carried out. Figure 5.11 shows the time traces of the temperature and streamwise velocity component at a fixed point ($x = 3$, $y = 3$, $z = 3$), which is the center of the computational domain, and the corresponding Fourier spectra are shown in Figure 5.12. It is clear that the two cases behave quite differently. For the nonswirling case, both the temperature

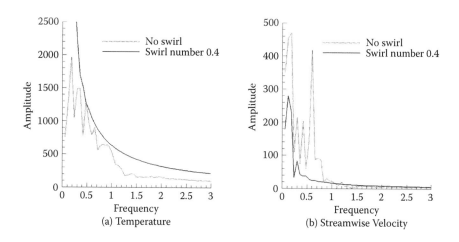

FIGURE 5.12   Fourier spectra of the temperature and streamwise velocity at ($x = 3$, $y = 3$, $z = 3$).

and streamwise velocity vary periodically, indicating that the flow field is dominated by the jet preferred mode of instability; the swirling case does not show similar behavior. Without swirl, the annular jet flow field is dominated by the unsteady vortical structures due to the Kelvin–Helmholtz instability. This instability is due to the existence of the two adjacent shear layers, which leads to the formation of vortical structures in the annular jet columns. Under swirling conditions, the vortical structures in the jet columns interact with the rotating inner structure. The presence of PVC in the swirling case changes the dynamics of the annular jet flame significantly. In Figure 5.11(a), it is observed that the temperature at $(x = 3, y = 3, z = 3)$ of the swirling case increases with time, while that of the nonswirling case is statistically more stable. In Figure 5.11(b), it can be seen that the streamwise velocity at $(x = 3, y = 3, z = 3)$ of the swirling case is close to a zero value, indicating a stationary flow field. With the presence of chemical reaction and combustion heat release at this location as indicated in Figure 5.9, the temperature shows an increasing trend that has not been stabilized for the time period shown because the heat has not been effectively removed. It is also noticed in Figure 5.11(b) that the velocity shows a slowly declining trend from zero value to negative values, which will bring the hot combustion product backward and thus will eventually stabilize the temperature.

An important feature in Figure 5.11 is that the reacting flow field without swirl at $(x = 3, y = 3, z = 3)$ is showing some sort of periodicity, while the flow periodicity is almost diminishing for the swirling case and the flow temperature is still changing significantly. The Fourier spectra shown in Figure 5.12 can further clarify this. In Figure 5.12(a) for the temperature, it can be seen that there are a few important frequencies in the lower frequency range in the nonswirling case, but the frequency of the excitation $f_0 = 0.30$ (nondimensional frequency – Strouhal number) is not particularly important. For the swirling case, the flow does not have important nonzero frequencies because the flow does not develop significant periodic behavior at $(x = 3, y = 3, z = 3)$. This is due to the fact that this location is within the inner structure of the swirl-induced large PVC characterized by rotating motions in the cross-streamwise direction and flow reversals in the streamwise direction, which does change significantly with time at this location. In Figure 5.12(b) for the streamwise velocity, two important frequencies have been observed: 0.18 and 0.60, where 0.60 is twice the frequency of the excitation. The development of the frequency 0.60 is mainly associated with the fact that two helical modes were superimposed, which not only broke the flow symmetry but also led to the development of the first super-harmonic frequency in the flow. For the swirling case, the flow velocity does not develop significant periodic behavior at ($x = 3$, $y = 3$, $z = 3$). The dynamic vortical structures in the annular jet columns interact with the inner structure of the swirl-induced large PVC, with the flow more or less stationary at the center of the domain. Unlike the dynamical vortical structures due to the Kelvin–Helmholtz type shear layer instability, the inner structure of the PVC does not respond to the flow instability supplied at the domain inlet; therefore, the flow periodicity is diminishing.

Fourier transformation is based on one-point history data and there is no correlation with neighboring points. Therefore, it may not reflect accurately the dynamics of the flow. To overcome this weakness, a POD analysis was performed for the history data of the flow field. As a powerful tool to investigate the mode effects in vortical flow fields and turbulence, POD can be used to analyze the flow data generated by DNS. The principle of POD is the decomposition of the flow field into a weighted linear sum of orthogonal eigenfunctions. The coherent structures in the flow field are described by the eigenfunctions of the two-point correlation tensor. The POD hypothesis is that different types of coherent motion that may occur within the flow will give rise to different POD eigenfunctions. The largest eigenvalue corresponds to the structure with most energy (Gunes and Rist 2004; 2007). The method of snapshots as described by Holmes et al. (1998) and Sirovic (1987) was utilized in this work to solve the associated eigenvalue problem.

In this method, an ensemble of M discrete instantaneous flow variables $\zeta(\vec{x}, t_k)$ (velocity fields in this study) acquired at time instants $t_k, k = 1, 2, \ldots, M$ is considered in a two-dimensional (2D) slice A of the 3D computational domain. The POD analysis is performed primarily in 2D slices cut in the streamwise and cross-streamwise directions to avoid the excessive computer memory requirements of a full 3D POD analysis of the DNS datasets. The full grid resolution of the DNS data is used. The time-average of the velocity field is computed and a new set of measurements $\xi(\vec{x}, t)$, which is the fluctuating velocity field, is calculated as follows:

$$\bar{\zeta}(\vec{x}) = \frac{1}{M} \sum_{k=1}^{M} \zeta(\vec{x}, t_k), \quad \text{and} \quad \xi(\vec{x}, t_k) = \zeta(\vec{x}, t_k) - \bar{\zeta}(\vec{x}) \tag{5.8}$$

A two-point correlation matrix C can be constructed as

$$C_{i,j} = \frac{1}{M} \int_A \xi(\vec{x}, t_i) \xi(\vec{x}, t_j) d\vec{x}, \quad \text{where } i, j = 1, 2, \ldots, M \tag{5.9}$$

The eigenvectors $\vec{a}_k^n$ and their corresponding eigenvalues λ_k can be found from the numerical solution of the equation

$$C \vec{a}_k^n = \lambda_k \vec{a}_k^n, \quad \text{where } k, n = 1, 2, \ldots, M \tag{5.10}$$

Using the eigenvectors $\vec{a}_k^n$ of matrix C, POD eigenfunctions $\phi^n(\vec{x})$ at mode n, which are optimal for the representation of the corresponding DNS data, can be linearly constructed by combining the fluctuating velocity as

$$\phi^n(\vec{x}) = \sum_{k=1}^{M} \vec{a}_k^n \xi(\vec{x}, t_k) \tag{5.11}$$

The POD eigenfunctions are orthogonal while the eigenvalues are positive $(\lambda_k \geq 0)$ in descending order $\lambda_k > \lambda_{k+1}$, where $k = 1, 2, \ldots, M$. Each eigenvalue quantifies the kinetic energy of the flow field datasets. The average fluctuating energy in the datasets can be calculated by summing up all the eigenvalues, $E = \Sigma_{k=1}^{M} \lambda_k$. The POD eigenfunctions can then be used to reconstruct the velocity fields as

$$\tilde{\xi}(\vec{x}, t) = \sum_{n=1}^{N} \vec{a}_k^n \phi^n(\vec{x}) \tag{5.12}$$

where N is the number of POD modes to be used for the reconstruction. Equation (5.12) is known as the "POD reconstruction formula." In general the first few modes capture most of the energy of the flow as quantified by the λ_k values. In other words, $N \ll M$ for flow reconstruction of large datasets using POD.

A POD analysis as described above was performed for the instantaneous velocity fields for the time interval between $t_1 = 30.00$ and $t_2 = 36.67$, where hundreds of instantaneous flow "snapshots" were recorded and analyzed. Sample results are shown in Tables 5.1 and 5.2 and Figures 5.13–5.16 for the velocity fields in the middle plane of the domain $y = 3$ for the two impinging cases. These results from the POD analysis correspond to

TABLE 5.1 Normalized Eigenvalues and Their Cumulative Contributions to the Fluctuating Energy in $y = 3$ Plane of the Nonswirling Case

Mode	1	2	3	4	5	6	7	8	9	10
λ_k, u component (%)	39.04	36.56	15.26	4.158	1.694	1.572	0.786	0.356	0.202	0.146
$\sum\lambda_k$, u component (%)	39.04	75.60	90.86	95.02	96.71	98.28	99.07	99.42	99.63	99.77
λ_k, w component (%)	38.17	26.76	20.07	7.886	2.488	2.155	1.271	0.456	0.263	0.196
$\sum\lambda_k$, w component (%)	38.17	64.93	85.01	92.89	95.38	97.53	98.81	99.26	99.52	99.72

TABLE 5.2 Normalized Eigenvalues and Their Cumulative Contributions to the Fluctuating Energy in $y = 3$ Plane of the Swirling Case

Mode	1	2	3	4	5	6	7	8	9	10
λ_k, u component (%)	37.00	31.66	17.63	7.893	1.799	1.464	1.034	0.578	0.361	0.176
$\sum\lambda_k$, u component (%)	37.00	68.67	86.30	94.19	95.99	97.46	98.49	99.07	99.43	99.61
λ_k, w component (%)	35.79	27.68	19.57	10.14	2.483	1.644	1.224	0.641	0.340	0.156
$\sum\lambda_k$, w component (%)	35.79	63.47	83.04	93.18	95.66	97.31	98.53	99.17	99.51	99.67

the last time instant $t_2 = 36.67$. Tables 5.1 and 5.2 provide the normalized eigenvalues and their cumulative contributions to the fluctuating energy in the plane of $y = 3$ for the nonswirling and swirling cases respectively. From Tables 5.1 and 5.2, it is clear that more than 95% of the total energy, which is the normal criterion used to judge the number of important modes in the flow field, can be captured in the first five POD modes in both computational cases, for both velocity components u and w. In the meantime, the first ten modes capture more than 99% of the total energy, whereas the rest of the modes contribute only a negligible amount of energy to the flow field. The trend of the mode effects is also shown in Figures 5.13 and 5.14, indicating an overall picture of the POD modal energy distributions in the flow field.

Figure 5.13 shows the energy content of each POD mode and fluctuating energy of the POD modes for the two velocity components (u, w) for the nonswirling case in the $y = 3$ plane, while Figure 5.14 shows those for the swirling case. In these figures, the "energy content" represents the contribution of an individual mode while the "fluctuating energy" represents the cumulative contributions of the relevant modes. From Figures 5.13 and 5.14, it is clear that the first four POD modes contain most of the energy of the flow. It is also observed that there is no significant difference between the modal energy distributions of the u component and w component. This indicates that the flow is of multidimensional nature. The formation of vortical structure due to the development of the Kelvin–Helmholtz instability in the flow field leads to a multidimensional flow field. There are several modes important to the energy distributions of the flow field, corresponding to the complex vortical flow field. The mode effects on the flow field are shown in Figures. 5.15 and 5.16.

POD analysis can be used to reconstruct the flow fields to illustrate the mode effects. Figures 5.15 and 5.16 show such reconstructions. In Figures 5.15, reconstructed velocity fields based on the first six most energetic POD modes are shown, in the section of $y = 3$ for the nonswirling case, while Figures 5.16 shows those for the swirling case. In both figures, it is evident that the flow field changes appreciably when the number of modes used in the velocity reconstruction is increased from 1 to 4, where the reconstructed flow field shows a gradual change with the increase in mode number. In the meantime, there is not any noticeable change when the number of modes used in the velocity reconstruction is increased from 4 to 6 while the reconstructed flow fields greatly resemble the corresponding

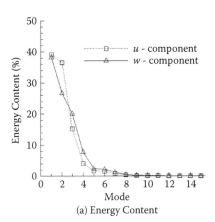
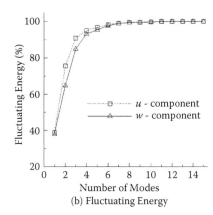

(a) Energy Content

(b) Fluctuating Energy

FIGURE 5.13   Energy content of each POD mode and fluctuating energy of the POD modes for the two velocity components (*u*, *w*) for the nonswirling case in the *y* = 3 plane.

DNS velocity field. In Figures 5.15 and 5.16, the differences between the velocity fields reconstructed using different numbers of modes indicate the amount of energy captured by each mode. The modal effects are evident in the flow field reconstructed using different numbers of modes, showing the flow structural development. As the number of modes increases, the

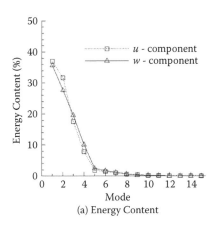
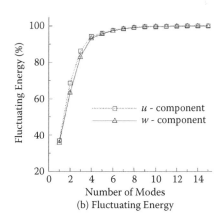

(a) Energy Content

(b) Fluctuating Energy

FIGURE 5.14   Energy content of each POD mode and fluctuating energy of the POD modes for the two velocity components (*u*, *w*) for the swirling case in the *y* = 3 plane.

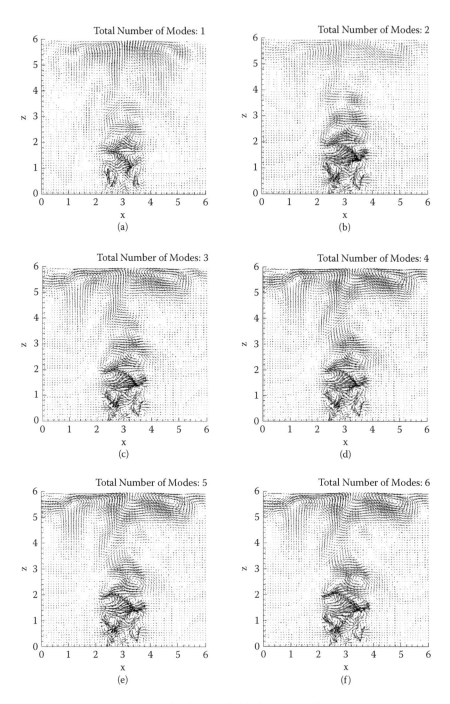

FIGURE 5.15   Reconstructed velocity fields based on the first six most energetic POD modes in $y = 3$ plane of the nonswirling case.

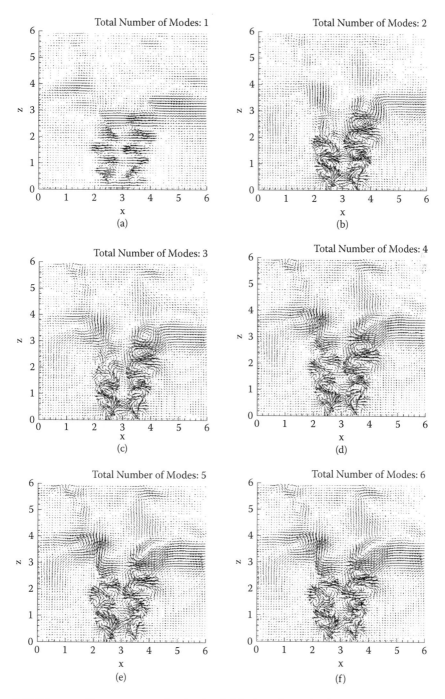

FIGURE 5.16   Reconstructed velocity fields based on the first six most energetic POD modes in $y = 3$ plane of the swirling case.

fluctuating energy captured increases and the reconstructed velocity field gradually approaches the DNS results.

In Figure 5.15, the most distinctive feature is that the wall effects are not captured when the number of modes used in the velocity reconstruction is small, indicating that the wall effects on the flow field are mainly associated with the higher mode numbers. In Figure 5.16, the most noticeable feature is that the first mode does not capture much of the flow velocity in the z direction, while it is recovered when the number of modes used in the velocity reconstruction is increased to 2. This indicates that the most energetic mode is mainly associated with the cross-streamwise velocity, while the second most energetic mode is associated with the streamwise velocity in this flow field. Clearly, POD analysis gives information on the dynamic feature of the flow field, which is not possible to obtain from a Fourier analysis.

The annular nonpremixed flame develops complex structures. Unsteady vortical structures in the jet column due to the Kelvin–Helmholtz–type shear layer instability originated from the two adjacent circular shear layers and have been observed at downstream locations. Under swirling conditions, the annular jet flame also develops PVC involving rotating motion on the inner side of the jet column. It was noticed that the PVC does not change appreciably with time, while the vortical structures in the jet column are highly unsteady. Time averaging of the results was also performed to examine the mean flow properties and the effects of swirl. The interval used for the averaging was $\Delta t: 30.00 \sim 36.67$, after the flow had reached a developed stage. Figure 5.17 shows the time-averaged flow and combustion properties along the centerline of the annular jet, while Figures 5.18 and 5.19 show the averaged reaction rate and streamwise velocity contours in the middle plane $y = 3$.

In Figure 5.17(a), it can be seen that the reaction rate profiles of both cases have a few peaks along the jet centerline. For the nonpremixed jet flame under investigation, the peaks observed in the averaged reaction rate profile correspond to locations where intense chemical reaction takes place when the fuel and oxidizer have been well mixed. In both cases, a small sharp peak located very close to the burner mouth can be observed. The small peak near the burner mouth is due to the formation of a recirculation zone near the fuel jet nozzle exit. In an annular configuration, the existence of the recirculation zone adjacent to the nozzle exit is associated with the formation of a stagnation region when the jet column meets at the center. The formation of this recirculation zone will bring the hot

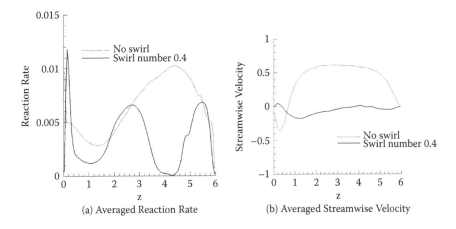

(a) Averaged Reaction Rate

(b) Averaged Streamwise Velocity

FIGURE 5.17   Time-averaged reaction rate and streamwise velocity profiles along the line ($x = 3$, $y = 3$).

combustion product back to and enhance the fuel/oxidizer mixing near the burner mouth, thus leading to the formation of a flame attached to the burner mouth. For the nonswirling case, a large peak at downstream locations is observed, which decays to zero value at the downstream wall boundary. For the swirling case, two large peaks are observed at the downstream locations. These large peaks can be better understood in conjunction with

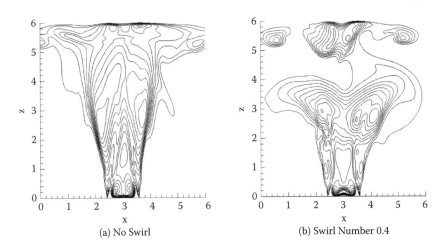

(a) No Swirl

(b) Swirl Number 0.4

FIGURE 5.18   Time-averaged reaction rate contours in the $y = 3$ plane (solid line: positive; dashed line: negative).

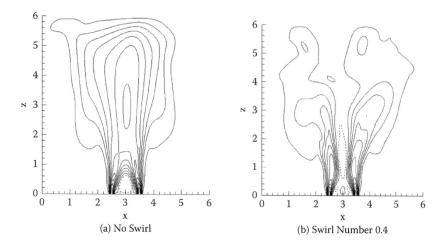

FIGURE 5.19 Time-averaged streamwise velocity contours in the $y = 3$ plane (solid line: positive; dashed line: negative).

the reaction rate contours shown in Figure 5.18. For the nonswirling case, the flame is continuous without intermittence. However, there is an intermittent region at the downstream location of the swirling flame, which is located above the "mushroom"-shaped flame. For the swirling case, the formation of reaction zones near the downstream wall above the intermittent region can still be observed, which is due to the enhanced fuel/oxidizer mixing associated with the wall impingement. The "mushroom"-shaped swirling flame indicates a shorter flame under swirling conditions. Swirl enhances the mixing; therefore, the flame length reduces under the swirling conditions but the unmixed fuel/oxidizer mixture can be further mixed and burnt in the downstream near-wall region due to the formation of a stagnation zone over there. For the annular swirling impinging flame, the interactions between the annular configuration, swirling motion, and wall impingement lead to several peaks in the averaged reaction rate along the jet centerline, as shown in Figure 5.17(a).

The annular jet flame also has a complex velocity field. The recirculation zone due to the annular configuration in the nonswirling case is located very close to the nozzle exit, the PVC of the swirling case is located at slightly downstream locations, and the stagnation zone due to wall impingement in both cases is located further downstream. Both the recirculation zone and the PVC involve flow reversals, indicated by the negative velocity values in Figure 5.17(b) and the dashed lines in Figure 5.19.

It is also noticed that the swirling case does not have appreciable stream-wise velocities at downstream locations after the PVC, while the nonswirl-ing case still develops significant streamwise velocities at downstream locations due to the continuous development of the flame. Under swirling conditions, the interaction between the recirculation zone associated with the annular configuration and the swirl-induced PVC affects the flow and flame dynamics. They are both located on the inner side of the jet column, involving flow reversals in proximity. Depending on the degree of swirl, the interaction may lead to different consequences: They may coexist or one zone may be overwhelmed by another. Figure 5.19 displays this inter-action. For the nonswirling case, only the "bell-shaped" recirculation zone associated with the annular configuration exists. For swirl number 0.4, the "bell-shaped" recirculation zone and the inner-structure of the PVC coex-ist, leading to a complex flow and an intensified reaction zone very close to the nozzle exit (indicated in Figure 5.17(a)) due to the enhanced mixing. As shown in Figures 5.17(b) and 5.19, the interaction between the recircu-lation zone and the PVC and the coexistence of them led to a small posi-tive streamwise velocity adjacent to the central point of the burn mouth.

From the averaged flow quantities shown, it is noticed that the vor-tical structures in the jet column due to the Kelvin–Helmholtz type shear layer instability cannot be observed because these structures are unsteady and are continuously convected downstream by the mean flow and would not appear when the flow field is time averaged. The results shown indicate that swirl has significant effects on the fluid dynamics of the flame because of the interaction between the recircu-lation zone near the jet nozzle exit and the PVC associated with swirl. The annular nonpremixed flame develops complex structures, where a recirculation zone is formed near the nozzle exit due to the annu-lar configuration that stabilizes the flame. Under swirling conditions, the annular jet also develops PVC involving recirculation and rotat-ing motion on the inner side of the jet column. Unlike the vortical structures in the jet column, which are highly unsteady, the recircula-tion zones do not change appreciably with time. At a moderate swirl number 0.4, the coexistence of the recirculation zone associated with the annular configuration and the inner PVC leads to a flame with strong reaction attached to the burner mouth. It was found that swirl has significant effects on flame dynamics. Under swirling conditions, the flame develops into a "mushroom"-like structure, and the flame becomes shorter but with much larger spreading at the downstream locations. The DNS results have revealed that the annular configuration stabilizes the flame, while swirl shortens flame length and enhances mixing and spreading of the flame.

Finally, the combustion results should be viewed with caution, bearing in mind the simplifications made in the DNS, for example, one-step global reaction, highly simplified transport properties, and relatively low Reynolds number. DNS with more realistic chemical and physical details will require substantially more CPU time than is readily available. However, given the fact that much combustion physics is yet to be explored in the interesting annular burner configuration, further DNS, LES, and experimental studies are justified.

II. NUMERICAL FEATURES: HIGH-ORDER SCHEMES FOR SPATIAL DISCRETIZATION

From the above examples, it can be seen that a common feature in the numerical methods used in DNS studies of compressible jet flows is that high-order finite difference schemes have been used for spatial differentiation. When periodic boundary conditions are employed, a spectral method is also used. Highly accurate numerical schemes are needed in DNS because turbulence cannot be resolved using lower-order numerical schemes where numerical diffusion can be larger than the small-scale turbulent transportation. The majority of existing DNS and LES codes for flow and combustion applications employ high-order finite difference schemes, due to the fact that their computing costs are generally lower than those of high-order finite volume methods (Ekaterinaris 2005). Although spectral methods are numerically accurate with low computing costs, they are not easy to apply for practical boundaries, complex domains, and compressible flows with discontinuities. Therefore, high-order finite difference schemes have been predominantly used in DNS of compressible flows for the spatial discretization. The time integration of the governing equations follows the time integration methods discussed in Chapter 3, while the specification of the boundary conditions follows those presented in Chapter 2. Therefore, they are not discussed here. The discussions of high-order finite difference schemes for spatial discretization closely follow those by Ekaterinaris (2005).

Finite difference schemes for spatial discretization can be obtained from Taylor series expansion. Continuous efforts have been made toward developing high-order finite difference (FD) methods for DNS and LES.

For solving the nonlinear Navier–Stokes equations, straightforward application of high-order accurate central difference schemes proved to be very problematic, due to the numerical instabilities from the spurious modes associated with the unresolvable high-frequency modes of the numerical discretization. Rai and Moin (1991) found that high-order upwind schemes are more promising to simulate turbulent flows. However, early attempts to apply high-order finite differences were often frustrating because of lack of robustness of the proposed high-order finite difference schemes compared to spectral methods. In spite of the difficulties, some success was achieved for the computation of incompressible flows (Rai and Moin 1991) and compressible flows (Rai and Moin 1993) with high-order, upwind FD schemes. For the simulation of rectangular jets, Rembold et al. (2002) used a fifth-order compact upwind-biased scheme for the spatial discretization of the convective terms. In Rai and Moin (1991), the fifth-order accurate derivatives were computed with upwind-biased formulae based on the sign of the velocity as

$$u_i = -\frac{1}{120}[6u_{i+2} + 60u_{i+1} + 40u_i - 120u_{i-1} + 30u_{i-2} - 4u_{i-3}] \quad \text{for } u_i > 0 \quad (5.13)$$

$$u_i = \frac{1}{120}[4u_{i+2} - 30u_{i+1} + 12u_i - 40u_{i-1} - 60u_{i-2} + 6u_{i-3}] \quad \text{for } u_i < 0 \quad (5.14)$$

Upwind-biased schemes alleviated some of the problems encountered with centered schemes such as numerical oscillations associated with the spurious modes. Upwind-biased schemes, however, based only on formal accuracy (truncation error) inherently introduce some form of artificial smoothing or dissipation error that makes them inappropriate for long-time integration such as that encountered in DNS and LES.

In contrast to upwind methods, central difference schemes do not introduce artificial dissipation. The dominant error in centered discretization is dispersion error, which is depressive. The stencils for the fourth-, sixth-, and eighth-order accurate symmetric, explicit, centered schemes are five-, seven- and nine-point wide, respectively. Large stencils are computationally disadvantageous because excessive amounts of matrix inversions are involved. As a result, only the fourth-order explicit scheme was used in CFD. Explicit fourth-order, finite difference formulae are often used to discretize the second-order derivatives in the viscous terms by taking the first derivative twice. In order to reduce the stencil width, the inner derivative is evaluated at half-points. A narrow stencil would be hugely advantageous for a central differencing with formal accuracy higher than sixth order.

The first systematic attempt to develop high-order accurate, narrow-stencil, finite difference schemes appropriate for problems with a wide range of scales was presented by Lele (1992). Compared to the traditional FD approximations, the compact schemes presented by Lele (1992) provided a better representation of the short-length scales. As a result, compact high-order schemes are closer to spectral methods and at the same time maintain the freedom to retain accuracy in complex stretched meshes. Emphasis on the development of compact schemes was given for the resolution characteristics of the difference approximations rather than formal accuracy or the truncation error. Compared with spectral methods, a significant advantage offered by the compact finite difference schemes is the convenience in the specification of boundary conditions. The centered compact or Padé schemes of Lele (1992) are briefly introduced as follows.

A. Centered Padé Schemes

The "compact" or Padé schemes presented by Lele (1992) can be derived from Taylor series expansions and they compute the derivatives simultaneously along an entire line in a coupled fashion. These centered schemes require small stencil support, which is of particular interest in DNS. The main advantage of compact schemes is simplicity in boundary condition treatment and smaller truncation error compared to their noncompact counterparts of equivalent order. The first-order derivative f' of a variable f in the governing equations can be obtained in the computational domain on an equally spaced mesh with size h, using a seven-point wide stencil finite difference discretization, given by

$$Bf'_{j-2} + Af'_{j-1} + f'_j + Af'_{j+1} + Bf'_{j+2}$$

$$= a\frac{f_{j+1} - f_{j-1}}{2h} + b\frac{f_{j+2} - f_{j-2}}{4h} + c\frac{f_{j+3} - f_{j-3}}{6h} \tag{5.15}$$

In Equation (5.15), coefficients A, B, a, b, and c determine the spatial accuracy of the discretization. The stencil becomes five-point wide if B and c are set to be zero. Different values of the coefficients in the formula yield

TABLE 5.3   Padé Schemes with Five- or Seven-Point Stencil

Scheme	E4	C4	C6	C8/5	C8/3	C10
A	0	$\frac{1}{4}$	$\frac{1}{3}$	$\frac{4}{9}$	$\frac{3}{8}$	$\frac{1}{2}$
B	0	0	0	$\frac{1}{36}$	0	$\frac{1}{20}$
a	$\frac{4}{3}$	$\frac{3}{2}$	$\frac{14}{9}$	$\frac{40}{27}$	$\frac{75}{48}$	$\frac{17}{12}$
b	$-\frac{1}{3}$	0	$\frac{1}{9}$	$\frac{25}{54}$	$\frac{1}{5}$	$\frac{101}{150}$
c	0	0	0	0	$-\frac{1}{80}$	$\frac{1}{100}$

schemes of different accuracy ranging from the fourth-order explicit method (E4) to the compact tenth-order accurate scheme (C10). The values of the coefficients in Equation (5.15) for schemes of different order of accuracy are shown in Table 5.3.

In this table, C8/3 refers to the eighth-order compact scheme that requires tridiagonal matrix inversion and C8/5 refers to the eight-order compact scheme that requires pentadiagonal matrix inversion. Apparently the pentadiagonal matrix inversion is more costly than the tridiagonal matrix inversion in terms of computational costs. For the tridiagonal matrix inversion, it can be conveniently achieved by the Gaussian elimination known as the tridiagonal matrix algorithm (TDMA), or Thomas algorithm (Conte and de Boor, 1972). The second-order derivative f'' of the variable f can also be obtained in the computational domain on an equally spaced mesh using a general form similar to the first-order derivative (Lele 1992). For both the first- and second-order derivatives, once the derivatives in the computational domain on an equally spaced mesh are obtained, derivatives in the physical domain with possibly nonuniform grid distribution can be obtained using the metrics for grid transformation.

The accuracy of different schemes can be assessed by the wavespace resolution of various explicit and compact schemes using Fourier analysis (Lele 1992). For Equation (5.15), considering that the exact result is a sinusoidal function $f_j = e^{ikh}$, where h is the uniform grid spacing and k is the wavenumber, the exact value of the derivative is $f_j' = ik f_j$, while the derivative computed with finite difference formula can be given by $\hat{f}_j' = i\hat{k} f_j$, where $\hat{k}$ is the modified wavenumber, which depends on the form of the FD formula used for the evaluation of the first-order derivative. The difference between the true wavenumber k and the modified wavenumber $\hat{k}$ is a measure of the scheme's resolving ability. The modified

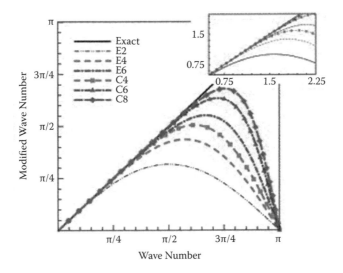

FIGURE 5.20 Wave space resolutions of explicit and compact centered schemes for the first-order derivative. (Ekaterinaris 2005; with permission from Elsevier Science Ltd.)

wavenumber of various finite difference schemes can be obtained using standard shift operators $\hat{f}'_{j\pm n} = \hat{f}'_j e^{\pm ik}$. For example, the modified number of the fourth-order accurate explicit scheme of Equation (5.15) is given by $\hat{k} = i[8\sin k - \sin(2k)]/6$, with analogous expression for the other methods. A comparison of the modified wavenumbers of the first derivative for several central compact and noncompact schemes is shown in Figure 5.20, using the scaled wavenumber $\omega = 2\pi kh/\lambda$, where λ is the wavelength and the number of intervals or grid points per wavelength is $2\pi k/\omega$. Therefore, the lower the scheme's resolving ability, the higher the number of points per wavelength required to resolve accurately a certain predetermined portion of the range $[0, 2\pi]$. This indicates that higher-order schemes need less grid points than do lower-order schemes for achieving the same level of accuracy.

As indicated by Equation (5.15) and Table 5.3, a seven-point stencil is needed for the eighth-order compact scheme C8/5 and the tenth-order scheme C10, which requires a pentadiagonal matrix inversion. In the meantime, a tridiagonal matrix inversion is needed for schemes such as the sixth-order scheme C6 and the fourth-order scheme C4. The tridiagonal matrix inversion can be conveniently achieved by the TDMA algorithm, which is less expensive than the pentadiagonal matrix inversion.

Therefore, the sixth-order and fourth-order compact schemes have been considered as good compromises between accuracy and computational speeds for DNS applications and have been broadly used in DNS codes. They are briefly summarized as follows. Arranging the coefficients in Equation (5.15) in a slightly different way, the first-order derivatives can be calculated using a five-point stencil from

$$f'_{j-1} + af'_j + f'_{j+1} = b\frac{f_{j+1} - f_{j-1}}{2h} + c\frac{f_{j+2} - f_{j-2}}{4h} \tag{5.16}$$

In Equation (5.16), the coefficients b and c can be calculated from a using $b = (2 + 4a)/3$ and $c = (4 - a)/3$ with $a = 4$ leading to a fourth-order scheme and $a = 3$ leading to a sixth-order scheme. Similarly, the second-order derivatives can be calculated using a five-point stencil from

$$f''_{j-1} + af''_j + f''_{j+1} = b\frac{f_{j+1} - 2f_j + f_{j-1}}{h^2} + c\frac{f_{j+2} - 2f_j + f_{j-2}}{4h^2} \tag{5.17}$$

In Equation (5.17), the coefficients are different from those in Equation (5.16), which can be calculated using $b = (4a - 4)/3$ and $c = (10 - a)/3$ with $a = 10$ leading to a fourth-order scheme and $a = 5.5$ leading to a sixth-order scheme.

B. Boundary Closures for High-Order Finite Difference Schemes

At the boundary points or points near the boundaries, central differencing is not possible because points outside the computational domain cannot be included. Inaccuracies in the application of discretization at boundaries can develop into spurious modes leading to numerical oscillations in the solution. One primary difficulty in using higher-order schemes is identification of stable boundary schemes that preserve their formal accuracy. Boundary closures for various explicit and compact high-order centered schemes were presented by Carpenter et al. (1993), who also assessed the stability characteristics of compact fourth- and sixth-order spatial operators with boundary closures.

At the boundary points or points near the boundaries, the discretization inevitably becomes one-sided. Numerical solutions of hyperbolic systems preserve their formal spatial accuracy when an nth-order inner scheme is closed with at least an $(n - 1)$th-order boundary scheme (Carpenter et al. 1993).

Considering the finite difference representation of the continuous derivative U_j' on an equally spaced mesh with the discrete form of the first-order derivative involving functional values at discrete points $j = 1, 2, \ldots, N$, for an explicit fourth-order accurate scheme with uniformly distributed mesh in the computational domain, the spatial discretization with boundary closures is obtained by Carpenter et al. (1993) as

$$
\left\{
\begin{aligned}
U_1' &= \frac{1}{12h}(-25U_1 + 48U_2 - 36U_3 + 16U_4 - 3U_5) \\[2mm]
U_2' &= \frac{1}{12h}(-3U_1 - 10U_2 + 18U_3 - 6U_4 + U_5) \\[2mm]
U_j' &= \frac{1}{12h}(U_{j-2} - 8U_{j-1} + 8U_{j+1} - U_{j-2}),\; j = 3, \ldots, N-2 \\[2mm]
U_{N-1}' &= \frac{1}{12h}(-U_{N-4} + 6U_{N-3} - 18U_{N-2} + 10U_{N-1} + 3U_N) \\[2mm]
U_N' &= \frac{1}{12h}(3U_{N-4} - 16U_{N-3} + 36U_{N-2} - 48U_{N-1} + 25U_N)
\end{aligned}
\right.
\tag{5.18}
$$

The fourth-order, compact FD scheme with narrower stencil for the approximation of U_j' requires closures only at $j = 1$ and $j = N$. The fourth-order compact scheme with boundary closures is

$$
\left\{
\begin{aligned}
U_1' + 3U_2' &= \frac{1}{6h}(-17U_1 + 9U_2 + 9U_3 - U_4) \\[2mm]
U_{j-1}' + 4U_j' + U_{j+1}' &= \frac{1}{h}(-3U_{j-1} + 3U_{j+1}),\; j = 2, \ldots, N-1 \\[2mm]
U_N' &= \frac{1}{6h}(U_{N-3} - 9U_{N-2} - 9U_{N-1} + 17U_N)
\end{aligned}
\right.
\tag{5.19}
$$

The sixth-order compact scheme has a five-point wide stencil and utilizes information from all five points explicitly and three points implicitly (tridiagonal system). Boundary closures must be provided at two points at each end of the domain $j = 1, 2$ and $j = N-1, N$. To ensure formal sixth-order formula, for example, the optimal scheme, in shorthand nomenclature, would be (5, 5-6-5, 5); for example, fifth order at the boundaries $j = 1, 2, j = N-1, N$, and sixth order in the interior. This closure preserves the formal spatial accuracy since an nth-order inner scheme is closed with an $(n - 1)$th-order boundary scheme, but a stable scheme at the boundary points was difficult to find (Carpenter et al. 1993). Therefore, the (3, 5-6-5, 3) and the (4, 5-6-5, 4) schemes are used for discretization with the sixth-order compact scheme.

The third-order closure at $j = 1$ is

$$2U_1' + 4U_2' = \frac{1}{h}(-5U_1 + 4U_2 + U_3)$$ (5.20)

The fourth-order closure at $j = 1$ is

$$6U_1' + 18U_2' = \frac{1}{h}(-17U_1 + 9U_2 + 9U_3 - U_4)$$ (5.21)

and the fifth-order closure at $j = 2$ is accomplished by

$$3U_1' + 18U_2' + 9U_3' = \frac{1}{h}(-10U_1 - 9U_2 + 18U_3 + U_4)$$ (5.22)

Practically, the (3, 4-6-4, 3) scheme is frequently used in DNS codes, referred to as the Padé 3/4/6 scheme in Chapter 2, where the formal accuracy of sixth order holds only in the interior of the computational domain. In this formulation, the number of neighboring points involved on one side of the boundary points or points near the boundaries is the same as that of the inner points. The scheme is of third-order accuracy at the boundary points, of fourth order at the next to the boundary points, and of sixth order at inner points only. The formula for the Padé 3/4/6 scheme were given in Equations (2.17)–(2.20). Although boundary closures are an essential part of high-order schemes, it was found that the effect of boundary closures on the overall resolution is indeed small even for highly accurate DNS (Adams 1998). Stable, accurate formula for the boundary points can be found in Carpenter et al. (1993).

Another issue encountered with high-order finite difference schemes used in CFD is grid uniformity, which affects the overall accuracy. The use of nonuniform grids in turbulent flow simulations is very often inevitable, especially in computational aeroacoustics where a large computational domain is needed (e.g., Jiang et al. 2004; 2006). The typical ratio of the maximum to the minimum grid spacing is about 100. The behavior of the second- and fourth-order explicit centered schemes and the fourth- and sixth-order compact schemes was assessed by Chung and Tucker (2003) for smoothly stretched grids. It was found that grid quality has stronger effects on the higher-order compact schemes than on the explicit schemes. Furthermore, an accuracy deterioration of higher-order compact schemes with low grid density was observed for nonuniform meshes.

C. Other High-Order Finite Difference Schemes

Mahesh (1998) presented a family of high-order finite difference schemes with good spectral resolution; they are more general than the standard compact schemes presented by Lele (1992). These schemes are symmetric and differ from the standard compact schemes in that the first and second derivatives are evaluated simultaneously. In addition, for the same stencil width, the schemes proposed by Mahesh (1998) are two orders higher in accuracy with significantly better spectral representation, and the computational cost for the evaluation of both derivatives is shown to be essentially the same as standard compact schemes. As a result, the proposed schemes appear to be attractive alternatives to standard compact schemes for the Navier–Stokes equations that include second-order derivative evaluation in the viscous terms. The schemes that compute simultaneously the first and the second derivatives of a function f given at a uniform mesh with spacing h are defined by

$$a_1 f'_{j-1} + a_0 f'_j + a_2 f'_{j+1} + h(b_1 f''_{j-1} + b_0 f''_j + b_2 f''_{j+1})$$

$$= \frac{1}{h}(c_1 f_{j-2} + c_2 f_{j-1} + c_0 f_j + c_3 f_{j+1} + c_4 f_{j+2}) \qquad (5.23)$$

By enforcing symmetry for the coefficients and considering $a_0 = 1$ and $b_0 = 1$, Mahesh (1998) obtained from Equation (5.23) a sixth- and an eighth-order compact scheme for the simultaneous evaluations of the first- and second-order derivatives. In the scheme of Mahesh (1998), the first and second derivatives are computed in a coupled fashion, using a narrower stencil compared to the standard compact scheme of Lele (1992).

High-order accurate centered schemes are nondissipative and they are particularly suitable for the convection of small-scale disturbances governed by the linearized equations such as the linearized Euler equations. Nondissipative, central-difference discretization for nonlinear problems, however, may produce high-frequency spurious modes that originate from mesh nonuniformities, inaccuracies of the boundary conditions, and nonlinear interactions. Spectral-type (Gaitonde and

Visbal 2000) or characteristic-based filters (Yee et al. 1999) may be used to stabilize numerical solutions performed with central-difference methods. Filtering of the solution with explicit-type filters was also proposed by Lele (1992). These filters are normally applied on the components of the computed solution vector. The characteristic-based filters such as those developed by Yee et al. (1999) remove spurious oscillations and in addition can be used for shock capturing. A recent improvement in characteristic-based filters is application of the essentially nonoscillatory (ENO) and weighted ENO (WENO) procedure in the evaluation of the dissipative fluxes (Garnier et al. 2001), which may be used for computations of highly compressible flows involving shock waves.

There are also other numerical schemes proposed for DNS of compressible flows. Sandham et al. (2002) proposed a stable high-order numerical scheme for DNS of shock-free compressible turbulence based on entropy splitting. The numerical scheme contains no upwinding, artificial dissipation, or filtering. Instead, the method relies on the stabilizing mechanisms of an appropriate conditioning of the governing equations and the use of compatible spatial difference operators in the interior scheme for the interior grid points as well as the in-the-boundary scheme for the boundary points. An entropy-splitting approach splits the inviscid flux derivatives into conservative and nonconservative portions. The spatial difference operators satisfy a summation-by-parts condition, leading to a stable scheme including both the interior and boundary points for the initial boundary value problem using a generalized energy estimate. A Laplacian formulation of the viscous and heat conduction terms on the right-hand side of the Navier–Stokes equations is used to ensure that any tendency to odd-even decoupling associated with central schemes can be countered by the fluid viscosity. The resulting methods are able to minimize the spurious high-frequency oscillations associated with pure central schemes, especially for long time integration applications such as DNS. For validation purposes, the methods were tested in a DNS of compressible turbulent plane channel flow at low values of friction Mach number, where reference turbulence databases existed (Coleman et al. 1995). It was demonstrated that the methods were robust in terms of grid resolution. Stability limits on the range of the splitting parameter were determined from numerical tests.

The high-order finite difference schemes have been a success in DNS and have been proven efficient and accurate. The symmetric, centered Padé schemes are nondissipative and favorable in terms of accuracy, but they produce nonphysical oscillations in the region of flow discontinuities such as shocks. This has limited the application of Padé schemes to relatively low-speed flows or high-speed flows without shock waves. For high-speed compressible flows containing shock waves such as those encountered in many aero-related problems, the numerical schemes need to be able to predict accurately the shock waves without nonphysical oscillations. To meet this challenge, there have been some recent efforts in combining a high-order Padé scheme with the ENO and WENO schemes (e.g., Wang and Huang 2002; Ren et al. 2003; Shen et al. 2006; Shen and Yang 2007). These hybrid Padé–ENO or Padé–WENO schemes have the advantages of the Padé scheme and ENO/WENO schemes. On the one hand, they possess the merit of the finite compact difference scheme, which is accurate and computationally efficient; on the other hand, they have the high-resolution property of ENO/WENO schemes for shock capturing.

For the computation of flows with discontinuities and nonlinearities, some form of upwinding is often needed. The idea of modifying or optimizing a finite difference scheme by calculating values of the coefficients that introduce upwinding or minimize a particular type of error instead of the truncation error has been used successfully in the design of new schemes with desired properties. Modifications of standard centered, explicit, and compact schemes were carried out by Zhong (1998). For the modified schemes, the formal order of the scheme for certain stencil size was sacrificed and high-order upwinding with low dissipation was introduced. Other optimized schemes have also been developed in the field of computational aeroacoustics. The rationale for optimizing numerical schemes for short waves is that for long waves, even lower-order schemes can do well. The short waves, however, require high resolution in order to obtain accurate representation of the broadband acoustic waves. For example, the optimized FD scheme of Tam and Webb (1993), referred to as the dispersion relation preserving (DRP) scheme, aims to predict accurately the short waves and uses central differences to approximate the first derivative. The approximation is therefore nondissipative in nature. The maximum formal order of accuracy of the centered scheme of Tam and Webb (1993) for certain stencil sizes is sacrificed in order to optimize resolution of the high wavenumbers. Although nondissipative schemes are ideal for aeroacoustics, numerical dissipation is often required to damp nonphysical waves generated by boundary and/or initial conditions. In many practical applications, therefore, high-order dissipative terms were added to the centered scheme of Tam and Webb (1993). There have also been optimized DRP schemes, such as those developed by Lockard et al. (1995) and Zhuang and Chen (1998), to alleviate the nonphysical waves associated with the boundary and/or initial conditions.

D. Finite Volume and Spectral-Volume Methods

Although the numerical methods used in DNS have been predominantly based on finite difference methods, there are a few severe problems such as incapability of dealing with complex geometries and difficulties in simulations of nonlinear phenomena. The main disadvantage of the finite volume (FV) formulation, especially when high-order accuracy is required, is the significantly higher computing cost of the FV methods compared to the finite difference (FD) formulation. The advantage of the FV formulation with respect to the FD formulation is that the former is based on the integral form of the conservation laws. As a result, flux conservation is enforced even on arbitrary meshes since the fluxes collapse telescopically by construction. Furthermore, analysis of the finite volume methods shows that they have superior performance in the high wavenumber range and that they exhibit lower truncation error, and therefore they are advantageous for the numerical simulation of nonlinear phenomena. Ekaterinaris (2005) presented the solution procedure of the multidimensional problem with the finite volume method, which consists of the following steps:

- Step 1. Reconstruction: Given the average values of the solution, reconstruct a polynomial approximation to the solution in each control volume. This polynomial may vary discontinuously from one control volume to another control volume. Reconstruction is the crucial step in the finite volume formulation.

- Step 2. Flux quadrature: Using the piecewise polynomial reconstruction of the solution, approximate the flux integral or in the discrete form the summation of fluxes by numerical quadrature.

- Step 3. Evolution and projection: A Riemann solver and an appropriate temporal discretization scheme are used to evolve the numerical approximation of the flux integral. (In FV methods, Riemann problems appear naturally for the solution of equations of conservation laws due to the discreteness of the grid.)

Gaitonde and Shang (1997) developed high-order (fourth- and sixth-order) compact difference-based schemes in the finite volume context. The formulation of these schemes utilizes the primitive function approach. Optimization of the schemes for better linear wave propagation characteristics can be performed by minimizing dispersion and isotropy errors.

Recently, a promising numerical method—the spectral-volume (SV) method (Wang 2002), which is another type of high-order accurate, conservative, and computationally efficient scheme—has been introduced to achieve high-order accuracy in an efficient manner similar to spectral methods and at the same time retain the benefits of the finite volume formulations for problems with discontinuities. In the spectral-volume method, cell-averaged data from each triangular or tetrahedral finite volume is used to reconstruct a high-order approximation in the spectral volume, while Riemann solvers are used to compute the fluxes at the spectral-volume boundaries. Since it does not require information from neighboring cells to perform reconstruction, it can be potentially very efficient and accurate. The spectral-volume method might be able to improve the capability of DNS and LES significantly since practical engineering problems are predominantly of complex geometry. However, for an SV method to be more broadly used, the computational efficiency and accuracy of the method with different orders of accuracy need to be systematically assessed and compared with the compact finite difference schemes with the same orders of accuracy.

REFERENCES

Adams, N.A. 1998. Direct numerical simulation of turbulent compression ramp flow. *Theoretical and Computational Fluid Dynamics* 12: 109–129.

Anderson, J.D. 2004. *Modern compressible flow with historical perspective.* New York: McGraw-Hill.

Boris, J.P. and Book, D.L. 1976. Flux-corrected transport. III. Minimal error FCT algorithms. *Journal of Computational Physics* 20: 397–431.

Canuto, C., Hussaini, M.Y., Quarteroni, A., and Zang, T.A. 1988. *Spectral methods in fluid dynamics.* New York: Springer Verlag.

Carpenter, M.H., Gottlieb D., and Abarbanel, S. 1993. The stability of numerical boundary treatments for compact high-order finite-difference schemes. *Journal of Computational Physics* 108: 272–295.

Chung, Y.M. and Tucker, P.G. 2003. Accuracy of higher-order finite-difference schemes on nonuniform grids, *AIAA Journal* 41: 1609–1611.

Colella, P. and Woodward, P. 1984. The piecewise parabolic method (PPM) for gas-dynamical simulations. *Journal of Computational Physics* 54: 174–201.

Coleman, G.N., Kim, J., and Moser, R. 1995. A numerical study of turbulent supersonic isothermal-wall channel flow. *Journal of Fluid Mechanics* 305: 159–183.

Conte, S.D. and de Boor, C.W. 1972. *Elementary numerical analysis*. New York: McGraw-Hill.

Duwig C. and Fuchs, L. 2005. Study of flame stabilization in a swirling combustor using a new flamelet formulation. *Combustion Science and Technology* 177: 1485–1510.

Ekaterinaris, J.A. 2005. High-order accurate, low numerical diffusion methods for aerodynamics. *Progress in Aerospace Sciences* 41: 192–300.

Gaitonde, D. and Shang, J.S. 1997. Optimized compact-difference-based finite-volume schemes for linear wave phenomena. *Journal of Computational Physics* 138: 617–643.

Gaitonde D.V. and Visbal M.R. 2000. Padé-type high-order boundary filters for the Navier–Stokes equations. *AIAA Journal* 38: 2103–2112.

García-Villalba, M. and Fröhlich, J. 2006. LES of a free annular swirling jet—Dependence of coherent structures on a pilot jet and the level of swirl. *International Journal of Heat and Fluid Flow* 27: 911–923.

García-Villalba, M. Fröhlich, J., and Rodi, W. 2006. Identification and analysis of coherent structures in the near field of a turbulent unconfined annular swirling jet using large eddy simulation. *Physics of Fluids* 18: 055103.

Garnier, E., Sagaut, P., and Deville, M. 2001. A class of explicit ENO filters with application to unsteady flows. *Journal of Computational Physics* 170: 184–204.

Grinstein, F.F. and Fureby, C. 2005. LES studies of the flow in a swirl gas combustor. *Proceedings of the Combustion Institute* 30: 1791–1798.

Gunes, H. and Rist, U. 2004. Proper orthogonal decomposition reconstruction of a transitional boundary layer with and without control. *Physics of Fluids* 16: 2763–2784.

Gunes, H. and Rist, U. 2007. Spatial resolution enhancement/smoothing of stereo-particle-image-velocimetry data using proper-orthogonal-decomposition-based and Kriging interpolation methods. *Physics of Fluids* 19: 064101.

Harten, A. 1983. High resolution schemes for hyperbolic conservation laws. *Journal of Computational Physics* 49: 357–393.

Harten, A., Engquist, B., Osher, S., and Chakravarthy, S.R. 1987. Uniformly high order accurate essentially non-oscillatory schemes. III. *Journal of Computational Physics* 71: 231–303.

Hirsch, C. 2007. *Numerical computation of internal and external flows*, vol. 2, 2nd ed. Oxford, UK: Butterworth-Heinemann.

Holmes, P., Lumley, J.L., and Berkooz, G. 1998. *Turbulence, coherent structures, dynamical systems and symmetry*. Cambridge, UK: Cambridge University Press.

Jakirlić, S. Hanjalić, K., and Tropea, C. 2002. Modeling rotating and swirling turbulent flows: a perpetual challenge. *AIAA Journal* 40: 1984–1996.

Jiang, X., Avital, E.J., and Luo, K.H. 2004. Sound generation by vortex pairing in subsonic axisymmetric jets. *AIAA Journal* 42: 241–248.

Jiang, X. and Luo, K. H. 2000. Direct numerical simulation of the puffing phenomenon of an axisymmetric thermal plume. *Theoretical and Computational Fluid Dynamics* 14: 55–74.

Jiang, X. and Luo, K. H. 2003. Dynamics and structure of transitional buoyant jet diffusion flames with sidewall effects. *Combustion and Flame* 133: 29–45.

Jiang, X., Siamas, G.A., and Wrobel, L.C., 2008. Analytical equilibrium swirling inflow conditions for computational fluid dynamics. *AIAA Journal* 46: 1015–1019.

Jiang, X., Zhao, H., and Cao, L. 2006. Numerical simulations of the flow and sound fields of a heated axisymmetric pulsating jet. *Computers & Mathematics with Applications* 51: 643–660.

Jiang, X., Zhao, H., and Luo, K.H. 2007. Direct numerical simulation of a non-premixed impinging jet flame. *ASME Journal of Heat Transfer* 129: 951–957.

Lele, S.K. 1992. Compact finite-difference schemes with spectral-like resolution. *Journal of Computational Physics* 103: 16–42.

Lockard, D.P., Brentner, K.S., and Atkins, H.L. 1995. High-accuracy algorithms for computational aeroacoustics. *AIAA Journal* 33: 246–253.

Mahesh, K. 1998. A family of high-order finite difference schemes with good spectral resolution. *Journal of Computational Physics* 145: 332–358.

Nichols, J.W., Schmid, P.J., and Riley, J.J. 2007. Self-sustained oscillations in variable-density round jets. *Journal of Fluid Mechanics* 582: 341–376.

Patte-Rouland, B., Lalizel, G., Moreau, J., and Rouland, E. 2001. Flow analysis of an annular jet by particle image velocimetry and proper orthogonal decomposition. *Measurement Science and Technology* 12: 1404–1412.

Poinsot T. and Lele, S.K. 1992. Boundary conditions for direct simulation of compressible viscous flows. *Journal of Computational Physics* 101: 104–129.

Poinsot, T. and Veynante, D. 2001. *Theoretical and numerical combustion.* Philadelphia: Edwards.

Rai, M.M. and Moin, P. 1991. Direct simulations of turbulent flow using finite-difference schemes. *Journal of Computational Physics* 96: 15–53.

Rai, M.M. and Moin, P. 1993. Direct numerical simulation of transition and turbulence in a spatially evolving boundary layer. *Journal of Computational Physics* 109: 169–192.

Reichert, R.S. and Biringen, S. 2007. Numerical simulation of compressible plane jets. *Mechanics Research Communication* 34: 249–259.

Rembold, B., Adams, N.A., and Kleiser, L. 2002. Direct numerical simulation of a transitional rectangular jet. *International Journal of Heat and Fluid Flow* 23: 547–553.

Ren, Y.X., Liu, M., and Zhang, H.X. 2003. A characteristic-wise hybrid compact-WENO scheme for solving hyperbolic conservation laws. *Journal of Computational Physics* 192: 365–386.

Roe, P.L. 1981. Approximate Riemann solvers, parameter vectors and difference schemes. *Journal of Computational Physics* 43: 357–372.

Sandham, N.D., Li, Q., and Yee, H.C. 2002. Entropy splitting for high-order numerical simulation of compressible turbulence. *Journal of Computational Physics* 178: 307–322.

Sankaran, V. and Menon, S. 2002. LES of spray combustion in swirling flows. *Journal of Turbulence* 3(1): 11.

Selle, L., Benoit, L., Poinsot, T., Nicoud, F., and Krebs, W. 2006. Joint use of compressible large-eddy simulation and Helmholtz solvers for the analysis of rotating modes in an industrial swirled burner. *Combustion and Flame* 145: 194–205.

Shen, Y.Q. and Yang, G.W. 2007. Hybrid finite compact-WENO schemes for shock calculation. *International Journal for Numerical Methods in Fluids* 53: 531–560.

Shen, Y.Q., Yang, G.W., and Gao, Z. 2006. High-resolution finite compact difference schemes for hyperbolic conservation laws. *Journal of Computational Physics* 216: 114–137.

Shu, C.-W. and Osher, S. 1988. Efficient implementation of essentially non-oscillatory shock capturing schemes. *Journal of Computational Physics* 77: 439–471.

Sirovic, L. 1987. Turbulence and the dynamics of coherent structures. *Quarterly of Applied Mathematics* XLV: 561–571.

Stein, O., Kempf, A.M., and Janicka, J. 2007. LES of the Sydney Swirl flame series: An initial investigation of the fluid dynamics. *Combustion Science and Technology* 179: 173–189.

Tam, C.K.W. and Webb, J.C. 1993. Dispersion-relation-preserving finite difference schemes for computational acoustics. *Journal of Computational Physics* 107: 262–281.

Uchiyama, T. 2004. Three-dimensional vortex simulation of bubble dispersion in excited round jet. *Chemical Engineering Science* 59: 1403–1413.

van Leer, B. 1979. Towards the ultimate conservative difference scheme. V. A second-order sequel to Godunov's method. *Journal of Computational Physics* 32: 101–136.

Vanoverberghe, K.P., Van den Bulck, E.V., and Tummers, M.J. 2003. Confined annular swirling jet combustion. *Combustion Science and Technology* 175: 545–578.

Wang, Z.J. 2002. Spectral (finite) volume method for conservation laws on unstructured grids—basic formulation. *Journal of Computational Physics* 178: 210–251.

Wang, Z.P. and Huang, G.P. 2002. An essentially nonoscillatory high-order Padé-type (ENO-Padé) scheme. *Journal of Computational Physics* 177: 37–58.

Yee, H.C., Sandham, N.D., and Djomehri, M.J. 1999. Low-dissipative high-order shock capturing methods using characteristic-based filters. *Journal of Computational Physics* 150: 199–238.

Zhong, X. 1998. High-order finite-difference schemes for numerical simulation of hypersonic boundary-layer transition. *Journal of Computational Physics* 144: 662–709.

Zhuang, M. and Chen, R.F. 1998. Optimized upwind dispersion-relation-preserving finite difference schemes for computational aeroacoustics. *AIAA Journal* 36: 2146–2148.

CHAPTER 6

LES of Incompressible Flows

A S AN ADVANCED CFD APPROACH, LES is very promising for practical applications. LES has been successfully applied to many industrial problems, in contrast to DNS, which has been mainly restricted to simple physical problems to understand the flow physics. LES has to be time-dependent three-dimensional calculations, similar to DNS. LES offers several advantages over the traditional CFD based on the RANS modeling approach and DNS. LES is able to predict instantaneous flow characteristics and resolve the large turbulent flow structures, in contrast to the traditional CFD based on the RANS approach, which provides only averaged flow quantities. Compared with DNS, the small scales are modeled in a LES approach and the requirement on accuracy of the numerical schemes is not as high as that of DNS. LES can be applied to problems with complex geometry using unstructured mesh and finite volume methods, but current DNS deals with only simple geometries using predominantly finite difference and spectral methods to achieve high-order accuracy. In terms of computational costs, LES is also much cheaper than DNS since very fine mesh is not required to resolve the small scales as in DNS.

This chapter is devoted to LES of incompressible flows, focusing on applications to complex geometries. First, some sample results are given, including results for both reacting and nonreacting flows. Second, subgrid scale (SGS) modeling, which is the most distinctive feature of LES, is briefly discussed in the context of incompressible flows. Finally, important numerical features are discussed, including LES on unstructured grids and the immersed boundary technique for complex geometries.

I. SAMPLE RESULTS: LES OF INCOMPRESSIBLE FLOWS IN COMPLEX GEOMETRIES

A. LES of Reacting Turbulent Flows in Complex Geometries (Mahesh et al. 2006; di Mare et al. 2004)

Mahesh et al. (2006) performed LES of a coaxial Pratt and Whitney gas-turbine combustor and an idealized coaxial dump combustor. The two geometries are shown in Figure 6.1. The incompressible Navier–Stokes equations are filtered to yield the incompressible LES equations. The sub-grid stress is modeled via a classical Smagorinsky model. A numerical algorithm that emphasizes discrete energy conservation on unstructured grid with hybrid arbitrary elements was used to solve the governing equations. The algorithm implies that the convective and pressure terms in the discrete kinetic energy equation may be expressed in divergence form. The integral of the kinetic energy over the computational domain is determined by its boundary fluxes and pressure-work at the boundaries. This is a useful property for the numerical algorithm since it implies that the sum of positive quantities is bounded. In the simulations performed by Mahesh et al. (2006), the numerical method was based on an earlier work

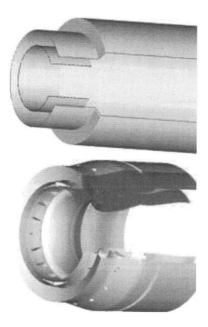

FIGURE 6.1   Schematics of an idealized coaxial dump combustor (top) and the combustor from a Pratt and Whitney gas-turbine engine (bottom). (Mahesh et al. 2006; with permission from ASME.)

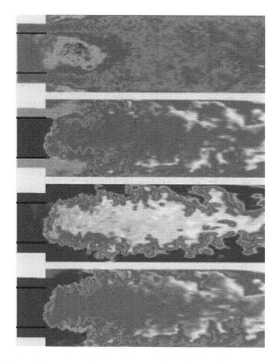

FIGURE 6.2 Instantaneous contours of streamwise velocity, temperature, CO mass fraction, and CO_2+H_2O mass fractions (from the top to
the bottom) of a gas-turbine engine. (Mahesh et al. 2006; with permission from ASME.)

by Mahesh et al. (2004), which is a predictor–corrector formulation that
emphasizes energy conservation for the convection and pressure terms
on arbitrary grids. In the method of Mahesh et al. (2004) for LES of complex geometry, time advancement is either explicit using the second-order
Adams–Bashforth method or fully implicit using the Crank–Nicholson
scheme along with linearization of the nonlinear terms. The Cartesian
components of momentum, density, and pressure are stored at the centroids of the computational elements, and the face-normal velocity is
stored at the centroids of the faces. The cell-centered momentum is predicted using the convective, viscous, and pressure-gradient terms at the
present time step. The predicted value of the momentum is then projected
such that the continuity equation is satisfied.

Figure 6.2 shows instantaneous contours of the velocity, temperature,
mixture fraction, and progress variable, respectively. Due to the high air/
fuel velocity ratio, a strong central recirculation region is formed in front of

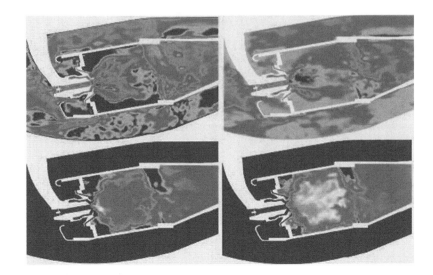

FIGURE 6.3 Instantaneous contours of velocity magnitude (top left), streamwise velocity (top right), mixture fraction (bottom left), and CO_2+H_2O mass fractions (bottom right) of a gas-turbine engine. (Mahesh et al. 2006; with permission from ASME.)

the fuel port. The recirculating combustion products provide a continuous ignition source for the relatively cold incoming reactants, thereby stabilizing the flame. Contours of the velocity magnitude, axial velocity, mixture fraction, and progress variable are shown in Figure 6.3. The correspondence between the progress variable and temperature is apparent. Swirl is observed to set up a recirculating region downstream of the injector. The fuel and air streams disperse radially and their interaction produces a lifted flame and an associated increase in temperature. The hot combustion products and recirculation flow sustain the flame, while the cold dilution jets interact with the combustion products to noticeably decrease the temperature. Figure 6.4 shows a snapshot of the region immediately downstream of the injector, where particle positions corresponding to the spray droplets are superposed on contours of velocity magnitude.

Figures 6.3 and 6.4 also indicate the level of geometric complexity; the combustor has numerous passages, holes of various sizes and shapes, swirlers, and obstacles in the flow path. The combustor chamber is fed by three coaxial swirlers and several dilution holes. The inlet air passes through the prediffuser and follows two paths; the main stream flows through the swirlers and enters the chamber, while the secondary stream is diverted to

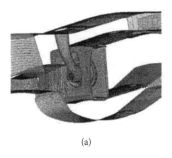
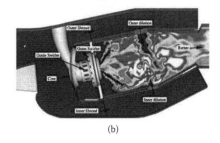

(a)   (b)

FIGURE 6.4   Gas turbine sector geometry (left) and instantaneous position of fuel spray superposed on contours of temperature from LES (right) of a gas-turbine engine. (Mahesh et al. 2006; with permission from ASME.)

the outer diffusers and enters the combustor through the dilution holes. The computations include the effects of flow bleed and transpiration, whose values are specified as boundary conditions. A hybrid unstructured grid of approximately 1.9 million hexahedral, tetrahedral, and pyramid elements was used to perform the simulations. After a statistically stable gas-phase flow was obtained, fuel was then respecified as liquid. The liquid fuel emerges as a conical spray, which breaks up and evaporates downstream of the injector. Clearly, the LES calculation of Mahesh et al. (2006) is a very complex one, involving the interactions between complex geometry, multiphase flow, and chemical reactions.

The reacting flows encountered in gas-turbine combustors have also been investigated by many other researchers using LES, for example, di Mare et al. (2004), who used LES to predict temperature and species concentrations in a model can-type gas-turbine combustor (Rolls-Royce Tay) operating in the nonpremixed combustion regime. Two computational cases were performed: Case I for a nondetailed swirl burner configuration and Case II for a detailed swirl burner configuration. Although the governing equations obtained by the density-weighted filter have a form similar to those of compressible flows, the solver is essentially an incompressible flow solver where the pressure was assumed to be thermodynamically constant based on the consideration that the Mach number is low in gas-turbine combustors. The time-dependent Navier–Stokes equations were solved for the Cartesian velocity components in a boundary conforming curvilinear coordinate system. A four-step second-order time-accurate approximate factorization method was applied to determine the pressure and ensure mass conservation in conjunction with a Rhie and Chow (1983) pressure smoothing technique to prevent even-odd node

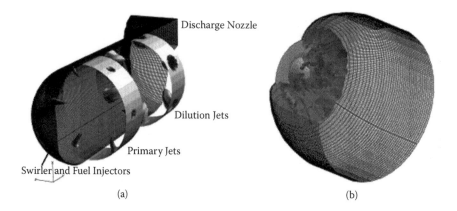

(a)   (b)

FIGURE 6.5   The computational mesh (left) and the fitted fuel injector grid (right) of a model gas-turbine combustor. (di Mare et al. 2004; with permission from Elsevier Science Ltd.)

uncoupling of the pressure and velocity fields. Standard second-order accurate numerical schemes in space and time were used. The subgrid scale stresses have been modeled by adopting the standard Smagorinsky–Lilly model. The computational mesh is shown in Figure 6.5, indicating the complex geometry of the physical problem.

Reynolds-averaged results were presented by di Mare et al. (2004). In Figures. 6.6 and 6.7, the differences between the flow topologies predicted in Cases I and II are particularly noticeable. In both instances it can be observed that the central core circulation interacts with the impinging primary jets, which drive the backflow in the primary region. A substantial amount of fluid is then pushed toward the wall, feeding a secondary

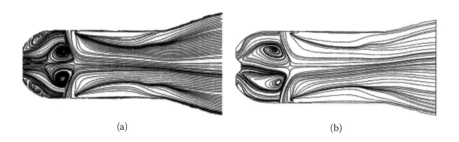

(a)   (b)

FIGURE 6.6   Reynolds averaged flow field: streamlines in the horizontal midplane of a model gas-turbine combustor, Case I (left) and Case II (right). (di Mare et al. 2004; with permission from Elsevier Science Ltd.)

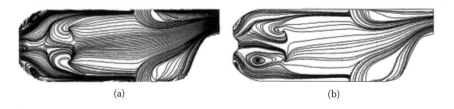

(a)   (b)

FIGURE 6.7 Reynolds averaged flow field: streamlines in the vertical mid-plane of a model gas-turbine combustor, Case I (left) and Case II (right). (di Mare et al. 2004; with permission from Elsevier Science Ltd.)

recirculation region. However, in Case I a smaller circulation is established above the fuel/air inlet. Its formation was believed to be associated with the approximate representation of the geometry whereby the 90° conical fueling device was replaced by a plain wall, thus approaching a bluff-body configuration. The presence of this additional toroidal circulation sensibly alters the flow configuration in the primary zone. The flow field downstream of the primary jets, however, is not influenced by the changes in the geometrical description of the fuel injector and the swirler exit conditions. In particular, the impingement region of the primary jets at the center of the combustor is clearly shown in Figure 6.8. The rotating core at the center of the combustor is an artifact of the averaging process as the location of jet impingement point varies with time. In Figure 6.8,

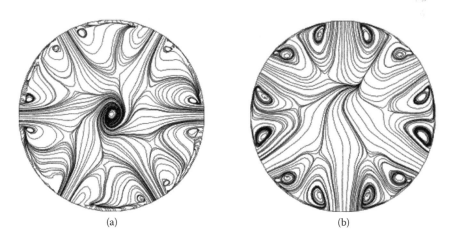

(a)   (b)

FIGURE 6.8 Reynolds averaged flow field of a model gas-turbine combustor, primary port cross-section (left) and dilution port cross-section (right) of Case I. (di Mare et al. 2004; with permission from Elsevier Science Ltd.)

a slight asymmetry in the penetration of the jets can also be identified. From the results of di Mare et al. (2004), it is evident that LES is able to reveal the complex flow patterns in gas-turbine combustors. In particular, di Mare et al. (2004) observed that an accurate description of the inlet section of the combustor plays a fundamental role in the prediction of the fuel placement and, hence, of the temperature distribution in the primary region of the gas turbine combustor.

B. LES of Separated Flows over a Backward-Facing Step (Dejoan and Leschziner 2004)

Separated flow behind a backward-facing step is a classical flow in CFD with practical relevance. In this flow configuration, the flow develops a separated shear layer that borders a recirculation zone behind the step. In the context of flow control using unsteady perturbations, Dejoan and Leschziner (2004) performed large-eddy simulation of periodically perturbed separated flows over a backward-facing step. The perturbation is provoked by the injection of a slot jet at zero-net-mass-flux, forming a synthetic jet and locating uniformly along the spanwise edge at which separation occurs. Due to the existence of this unsteady perturbation, the traditional CFD based on the RANS modeling approach was regarded as inappropriate. In LES performed by Dejoan and Leschziner (2004), attention focused on one jet-forcing frequency, at a nondimensional frequency or Strouhal number of 0.2, for which experimental data showed the perturbation to cause the largest reduction in the time-mean recirculation length.

The geometry of the flow simulated is shown in Figure 6.9. In the simulations performed, the inlet stream was turbulent and fully developed at the step with a Reynolds number $Re = 3700$, which was based on the maximum inlet velocity and step height. Two flows were simulated: one without injection and the other one with a slot jet that is injected at the step at $45°$ into the flow across the entire span through a 1-mm slit at a

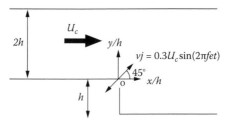

FIGURE 6.9 Schematic of the backward-facing-step flow. (Dejoan and Leschziner 2004; with permission from Elsevier Science Ltd.)

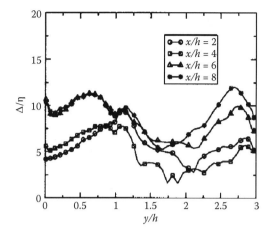

FIGURE 6.10   Ratio of the cell size to the Kolmogorov scale along various grid lines for the unperturbed flow. (Dejoan and Leschziner 2004; with permission from Elsevier Science Ltd.)

sinusoidal velocity, as indicated in Figure 6.9. The inlet conditions were generated by a precursor simulation for a fully developed channel flow in a periodic domain at the appropriate Reynolds number. At the outlet, the convective outflow condition was used. The domain downstream of the step is covered by a mesh of $96 \times 80 \times 32$ cells, while the inlet channel is covered by a separate mesh of $96 \times 40 \times 32$ cells, leading to the total number of cells of around 0.4 million with fine grid located in the step and injection region. The mesh size used in the simulation was usefully analyzed in terms of wall unit and the Kolmogorov length scale by Dejoan and Leschziner (2004). The analysis showed that the cell-aspect ratios are, typically, $\Delta y^+/\Delta x^+/\Delta z^+ = 1.5/28/20$ at the wall and $4.5/28/20$ in the shear layer. Figure 6.10 shows the ratio between a mesh size indicator $\Delta = (\Delta x \Delta y \Delta z)^{1/3}$ and the Kolmogorov length scale $\eta = (\nu^3/\varepsilon)^{1/4}$ along several constant x/h and y/h lines, where the dissipation rate used in the estimation of the Kolmogorov length scale was obtained from the turbulence-energy budget. This analysis showed that the cutoff of the subgrid scales lies close to the dissipative part of the wave-number range. Subgrid scale processes were represented by means of either the Smagorinsky model with van Driest damping or the wall-adapting local-eddy viscosity (WALE) model of Ducros et al. (1998). Significant differences between the statistical properties of the solutions including the second moments were not observed from these two SGS models.

In the LES calculations performed by Dejoan and Leschziner (2004), the computational scheme used is a general multiblock finite volume procedure with nonorthogonal-mesh capabilities. The finite volume scheme is second-order accurate in space, using central differencing for advection and diffusion. Time-marching is based on a fractional-step method, with the time derivative being discretized by a second-order backward-biased approximation. The flux terms are advanced explicitly using the Adams–Bashforth method. The provisional velocity field is then corrected via the pressure gradient by a projection onto a divergence-free velocity field. In the solution procedure, the pressure is computed as a solution to the pressure-Poisson problem by means of a partial-diagonalization technique as described by Schumann and Sweet (1988) and a multigrid algorithm operating in conjunction with a successive-line over-relaxation scheme.

Dejoan and Leschziner (2004) reported results for time-averaged and phase-averaged velocity and Reynolds stresses, obtained from their LES calculations. Profiles of Reynolds stresses are shown in Figure 6.11,

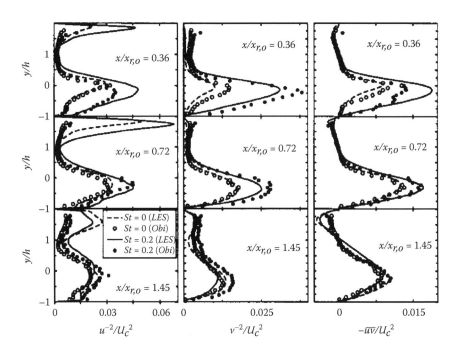

FIGURE 6.11 Time-averaged Reynolds stresses at different streamwise locations downstream of the step for the unperturbed and perturbed flows. (Dejoan and Leschziner 2004; with permission from Elsevier Science Ltd.)

together with the corresponding experimental data for comparison. The most striking feature in the Reynolds stress profiles is the significant increase in turbulence activity provoked by the injection in the shear layer: close to the step, the lateral normal stress has increased by a factor 3, while the shear stress has risen by a factor 2. These increases are, qualitatively, consistent with the substantial reduction in recirculation length, since the injection provokes a much higher transport of momentum and hence more rapid recovery. Another striking feature is the substantial increase in the turbulence activity within the recirculation zone, due to the formation of significant unsteady features below the shear layer associated with the "flapping" of that layer, which can be identified from the phase-averaged results. In addition to the time-averaged results, an overall view of the flow periodic behavior is indicated in Figure 6.12, which shows contours of the phase-averaged velocity and pressure fluctuations at four phase locations separated by $\pi/2$. Both the stream function and pressure fields show that periodicity in the perturbation produces large-scale vortices below the separated shear layer. The overall picture thus presented is one in which the shear layer is "flapping," a process related to the interaction of the large-scale vortical

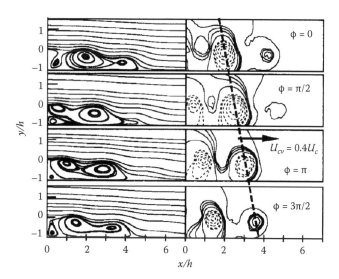

FIGURE 6.12 Phase-averaged stream function and phase-averaged pressure fields in the reattachment region with negative pressure contours identified by dashed lines and positive pressure contours identified by solid lines. (Dejoan and Leschziner 2004; with permission from Elsevier Science Ltd.)

structures with the wall around the reattachment zone. The simulations provide clear indications that the high level of sensitivity to the perturbation at the Strouhal number considered is due to a strong interaction between shear-layer instabilities and shedding-type instabilities induced by the interaction of large-scale structures developed downstream of the step with the wall, causing the shear layer to flap. The LES calculations by Dejoan and Leschziner (2004) clearly indicate that LES is capable of revealing the physics of flow unsteadiness.

II. SUBGRID SCALE MODELING OF INCOMPRESSIBLE FLOWS

A. Subgrid Scale Modeling

Subgrid scale (SGS) modeling is the most distinctive feature of LES. The SGS Reynolds stress in LES is due to the filtering or local average of the flow field, unlike the Reynolds stress in RANS, which is due to a time or ensemble average. When a relatively fine mesh is used, the SGS kinetic energy is a small part of the total flow kinetic energy; therefore, model accuracy becomes less crucial in LES than in RANS computations where the turbulent energy can be a significant part of the total flow energy and does not depend strongly on the mesh used in the computation. In LES, the governing equations are spatially filtered rather than time or ensemble averaged. Explicit account is taken of flow structures larger than the filter width, while the influence of unresolved scales is modeled. In LES, it is essential to define the quantities to be computed precisely as in the RANS approach. To do this it is essential to define a field that contains only the large-scale components of the total field. In LES, a spatial filter is applied to the equations of motion so that motions with scales less than the filter width do not have to be resolved by the computational mesh. The spatial filter of a function $f = f(\mathbf{x},t)$ is defined as its convolution with a filter function, G, according to

$$\bar{f} = \int_{\Omega} G(\mathbf{x} - \mathbf{x}'; \Delta(\mathbf{x})) f(\mathbf{x}',t) d^3\mathbf{x}' \qquad (6.1)$$

In Equation (6.1), the integration is carried out over the entire flow domain Ω. The filter function has a width Δ, which is a function of the three-dimensional location $\mathbf{x}$ and may vary with position. A number of different filter kernels $G(\mathbf{x} - \mathbf{x}'; \Delta(\mathbf{x}))$ may be used, in the form of a localized function that is large only when $\mathbf{x}$ and $\mathbf{x}'$ are not far apart. When the Navier–Stokes equations are filtered, one obtains a set of equations very similar in form to the RANS equations. For an incompressible flow, the filtered Navier–Stokes equations used in LES contain a term $\overline{u_i u_j}$. Since $\overline{u_i u_j} \neq \overline{u}_i \overline{u}_j$, a modeling approximation for the difference between the two sides of this inequality $\tau_{ij} = \overline{u_i u_j} - \overline{u}_i \overline{u}_j$ must be introduced. In the context of LES, $\tau_{ij} = \overline{u_i u_j} - \overline{u}_i \overline{u}_j$ is called the SGS Reynolds stress. It plays a role in LES similar to the role played by the Reynolds stress in RANS models but the physics that it models is different. Similar to the unknown Reynolds stress in RANS approach, the SGS Reynolds stress is an unknown and needs to be modeled using SGS models so that the governing equations are a closed set.

The justification for LES is that the larger eddies contain most of the energy, do most of the transporting of conserved properties, and vary most from flow to flow; the smaller eddies are believed to be more universal and less important and should be easier to model. It is hoped that the SGS Reynolds stress can be more readily modeled due to the universality involved. In RANS modeling, however, the Reynolds stress can vary significantly from flow to flow and from one part of the flow field to another, which could lead to severe modeling inaccuracies. The SGS Reynolds stress in LES and the Reynolds stress in RANS are physically and numerically different. The SGS Reynolds stress in LES is due to a local average or filtering of the complete field, while the Reynolds stress in RANS is due to a time or ensemble average. In most the cases, the SGS energy is a much smaller part of the total flow than the RANS turbulent energy so model accuracy may be less crucial in LES than in RANS computation.

As described in Chapter 1, the most commonly used subgrid scale model is the one proposed by Smagorinsky (1963), based on the Boussinesq hypothesis or assumption (Boussinesq 1877), which marked the beginning of LES. It is an eddy viscosity model, given as

$$\tau_{ij} - \frac{1}{3}\tau_{kk}\delta_{ij} = -\nu_T\left(\frac{\partial \overline{u}_i}{\partial x_j} + \frac{\partial \overline{u}_j}{\partial x_i}\right) = -2\nu_T \overline{S}_{ij} \tag{6.2}$$

where $\overline{S}_{ij}$ is the resolved strain rate tensor, δ_{ij} is the Kronecker delta, and ν_T is the eddy viscosity. The well-known Smagorinsky–Lilly model is the direct equivalent of Prandtl's mixing length model used in statistical turbulence modeling. The natural choice in LES for the length scale is the filter width Δ, and the corresponding choice for the velocity scale is $\Delta|\overline{S}|$, where $\overline{S} = \sqrt{2S_{ij}S_{ij}}$ is a measure of the velocity gradient. Correspondingly, the eddy viscosity in Equation (6.2) can be given by

$$v_T = (C_S\Delta)^2 \,|\overline{S}| \qquad (6.3)$$

where C_S is the so-called Smagorinsky constant. As discussed by Geurts (2004), $C_S = 0.1$ is the most commonly used value although a value of $C_S = 0.17$ was initially put forward for homogeneous isotropic turbulence. In a three-dimensional LES, the filter width is often determined by the local mesh size as $\Delta = (\Delta x\,\Delta y\,\Delta z)^{1/3}$ for a Cartesian grid.

In the Smagorinsky model, both the length scale and velocity scale are prespecified. The model does not take into account the local flow properties, which may cause problems in a complex geometry domain where the flow may experience significant spatial changes. To allow a closer adaptation of the eddy viscosity to local flow properties, the analogue of the one- or two-equation models used in the RANS approach or the statistical turbulence modeling may be considered, as discussed by Ghosal et al. (1995). For instance, in one-equation models, only the length scale, which can be conveniently taken as the filter width in LES, is prespecified and the velocity scale can be taken as $\sqrt{k}$ where the turbulent kinetic energy k is determined from a separate transport equation. Such a k-equation model was also proposed by Menon et al. (1996) and used by several researchers (e.g., Valentino et al. 2007) in the LES of transient gas jets and sprays under diesel conditions. In a k-equation SGS model, the eddy viscosity can be given by

$$v_T = C\sqrt{k}\,\Delta \qquad (6.4)$$

In Equation (6.4), C is a model constant.

Since most of the practical problems are of complex geometry, SGS models for complex geometries are of particular importance. Over the last two decades, advances have been made in SGS modeling that are particularly appropriate for LES in complex geometries. Complex flows usually contain multiple flow regimes such as wall boundary layers and flow separations. It has been demonstrated that models with a fixed coefficient require tuning of their coefficients in each flow regime. The dynamic modeling approach (e.g., Germano et al. 1991; Moin et al. 1991; Ghosal et al. 1995) does not suffer from this limitation because the model coefficient is a function of space and time, and is computed rather than prescribed. In addition, it has the proper limiting behavior near walls without ad hoc damping functions and behaves appropriately in the transition regions. These are all very important features for LES in complex domains. There are other SGS models intended for LES for complex geometries; for example, Domaradzki and Loh (1999) used extrapolation from the resolved scales to subgrid scales to construct the subgrid scale velocity fluctuations and stresses. The model has an adjustable parameter that should be possible to compute dynamically, thereby making the model suitable for complex geometries. Hughes et al. (2001) have also shown that better results are obtained if the governing equations for LES are split into the large- and small-scale equations, and the eddy-viscosity model is applied only to the small-scale equations. The extension of this approach to complex geometry appears to be straightforward with a variational formulation as proposed by Hughes et al. (2001).

The dynamic Smagorinsky model as discussed by Germano et al. (1991) and Moin et al. (1991) is useful to LES of complex geometries. It can be extended to unstructured grids, which was used in the LES calculations by Mahesh et al. (2006). For the eddy viscosity given in Equation (6.3), application of the dynamic procedure using the least-squares approach described by Lilly (1992) yields the following expression:

$$(C_s\Delta)^2 = -\frac{1}{2}\frac{L_{ij}M_{ij}}{M_{kl}M_{kl}} \text{ with } L_{ij} = \widehat{\bar{u}_i\bar{u}_j} - \hat{\bar{u}}_i\hat{\bar{u}}_j, \; M_{ij} = \left(\frac{\hat{\Delta}}{\Delta}\right)^2\widehat{|\bar{S}|\bar{S}_{ij}} - |\hat{\bar{S}}|\widehat{\bar{S}}_{ij} \quad (6.5)$$

In Equation (6.5), ^ represents a test filter, which is required by the dynamic procedure, and the ratio of test to grid filter widths $\hat{\Delta}/\Delta$ is commonly assumed to be 2 (Mahesh et al. 2004). In a finite volume approach, the filter width can be defined as $\Delta = V^{1/3}$ with V denoting the element volume, which yields a filter width of $\Delta = (\Delta x\,\Delta y\,\Delta z)^{1/3}$ for a Cartesian grid.

Near-wall flows are always encountered in complex geometries and are difficult for the SGS model as the flow may be laminar and transitional and may develop deterministic near-wall structures. Ducros et al. (1998) proposed a wall-adapting local eddy-viscosity (WALE) model, which was used in the LES calculations performed by Dejoan and Leschziner (2004). Compared to the standard Smagorinsky model, the WALE model reproduces the cubic wall-asymptotic behavior of the subgrid scale viscosity and gives lower values of this viscosity. Nevertheless, SGS modeling of near-wall flows is particularly challenging since SGS models were not designed to account for the highly deterministic near-wall structures. As discussed in Chapter 2, the RANS approach in the near-wall region is often adopted, leading to hybrid LES–RANS approaches. There were efforts to develop a dynamic approach to the near-wall flow modeling. For instance, Wang and Moin (2002) used the RANS approach in the near-wall region of Balaras et al. (1996), but incorporated a new dynamic approach to adjust the model coefficient. The basic rationale for the adjustment is that when a RANS-type eddy viscosity is used in the wall layer equations, which includes non-linear convective terms, its value must be reduced so that the eddy viscosity would account for only the unresolved part of the Reynolds shear stress. In general, a dynamic approach is expected to perform better for complex geometries than an SGS model with a fixed model coefficient.

B. SGS Modeling of Reacting Flows

SGS models for complex flows such as reacting flows encountered in combustion applications are more complex than those for nonreacting flows. In spite of the huge success of LES in predicting nonreacting flow fields, LES of reacting flows is much more challenging. One important reason for this is that the chemical reactions in reacting flows may occur at the smallest dissipative scales controlled by molecular diffusion. In other words, the combustion process occurs at the subgrid level, which has to be modeled entirely in LES by the SGS model. This is significantly different from the LES of nonreacting flows, where the SGS model represents only a small part of the energy containing scales while the major part of the energy containing scales are resolved. In a reacting flow, combustion significantly changes the fluid dynamic behavior due to the change in temperature, fluid viscosity, density, and fluid composition. In combustion applications, chemical reaction and the associated heat release introduce fine-scale density and velocity fluctuations that in turn couple the small scale events back to the larger fluid-dynamical scales.

There has been a substantial amount of work devoted to LES of reacting flows, where the SGS model depends on how the combustion is mathematically represented. In principle, the system of governing equations for reacting flows could include the transport equations for chemical species, according to the assigned reaction mechanism, and energy equation, in addition to the Navier–Stokes momentum equations. Compared with the governing equations for nonreacting flows, there is not only a dramatic increase in the number of equations to be solved, but also presence of nonlinear filtered chemical source terms that must be modeled. LES of reacting flows is therefore a difficult and computationally expensive task. However, various assumptions can be used to simplify the mathematical formulation and combustion modeling. For example, with a number of fairly benign assumptions—the heat-releasing chemical reaction is infinitely "fast," the flow is adiabatic, the Prandtl and Schmidt numbers are equal, the pressure is thermodynamically constant, etc.—the instantaneous major species composition, temperature, and density can be related to a strictly conserved scalar quantity, the mixture fraction, as described by di Mare et al. (2004). In this way, compared with a nonreacting flow system, only one additional equation needs to be solved for the reacting flow system, which is the governing equation for the mixture fraction containing no reaction source terms. The filtered governing equation for the mixture fraction contains an unknown subgrid scalar flux, but can be modeled using a gradient transport model involving the subgrid Prandtl/Schmidt number. This is perhaps the simplest LES formulation and SGS modeling for reacting flows.

Compared to turbulent time scales, combustion introduces faster time scales, which makes the modeling of reacting flows more challenging than that of nonreacting flows. For reacting flows, the multiple time scale phenomenon needs to be modeled appropriately. One modeling effort applied to LES of combustion, the linear eddy mixing (LEM) model as discussed by Menon et al. (1993), is particularly interesting in the context of SGS modeling of reacting flows. The LEM model can be applied to both incompressible and compressible flows, as discussed by Chakravarthy and Menon (2001) and Sankaran and Menon (2005). In the LEM model, the physical processes in turbulent combustion are considered as large-scale advection and subgrid scale mixing, molecular diffusion, and chemical reaction, which can be incorporated together without incurring high computational costs. LEM is essentially a two-scale approach. In the LEM model, the scalar fields are not spatially filtered along with the other equations. Instead, a subgrid Eulerian method is combined with a Lagrangian LES transport method to evolve the unfiltered scalar fields. Subsequently, the LES-resolved filtered species are obtained by ensemble averaging the subgrid fields. The numerical implementation of LEM employs a subgrid mixing and combustion model within each LES cell in which molecular diffusion and kinetics are simulated exactly in a spatial dimension that is oriented in the direction of the maximum scalar gradient using a resolution fine enough to resolve all the scales in the local LES cell. Subgrid stirring by eddies smaller than the grid is implemented as a series of stochastic processes, each of which represents the action of a turbulent eddy on the scalar field. Heat release in the subgrid LEM domain results in volumetric expansion that is included as an increase in the subgrid volume. As discussed by El-Asrag and Menon (2007), there are several important features in the LEM model. One such feature is that molecular diffusion including possible differential diffusion at the small scales is simulated exactly in the LEM domain, albeit in the direction of the maximum scalar gradient, which is the most critical direction. In a LEM model, finite-rate kinetics can also be included without requiring any closure. In addition, the effects of turbulent mixing on the subgrid scalar field due to subgrid stirring by small-scale eddies, although implemented in a stochastic sense, are explicitly incorporated. Within the context of LES of reacting flows, the LEM approach can be used to model the small-scale processes ranging from the grid resolution down to the Kolmogorov scale or the smallest scales related to chemical reaction in reduced dimension, while the large scales of the flow are calculated directly from the LES equations of the motion with an appropriate coupling procedure.

LES of reacting flows is an area requiring continuous efforts. SGS models for reacting flows are often complicated by the involvement of multiphase flow phenomena, due to the fact that combustion always utilizes liquid fuels such as petrol or diesel while the flame is a gas-phase phenomenon. SGS models for complex multiphase flows are very immature. There is a lack of well-established SGS models, especially for the interactions between the different phases. There is no SGS model available to date that could take into account the subgrid influence of the dispersed phase that is locally smaller than the grid size, such as the fine liquid droplets or solid particles dispersed in a gas medium, on the resolved scales in the gas phase. Clearly more efforts are needed in the area of SGS modeling of multiphase reacting flows.

III. NUMERICAL FEATURES: LES ON UNSTRUCTURED GRIDS AND IMMERSED BOUNDARY TECHNIQUE FOR COMPLEX GEOMETRIES

On the numerical aspects, there are similarities and differences between DNS and LES. As discussed in Chapter 1, both DNS and LES have to be time-dependent three-dimensional simulations; therefore, they share much commonality in the numerical treatment of boundary conditions and numerical methods for time integration, as discussed in Chapters 2 and 3. However, higher-order schemes are more often required in DNS than in LES, both for time integration and spatial discretization. In the state-of-the-art DNS, the discretization schemes used are at least fourth order, typically sixth and above. In LES, the numerical schemes used are normally between second and fourth order. In order to achieve higher order of the numerical scheme, finite difference and spectral methods are frequently used in DNS. This, however, is achieved at the expense of the efficacy in dealing with complex geometries. Consequently, DNS has been mainly restricted to simple geometry problems to understand the flow physics. On the contrary, the requirement on the numerical accuracy in LES can be more relaxed than in DNS, due to the fact that fine scales do not need to be resolved in LES. This not only leads to the wide use of slightly lower-order numerical schemes in LES such as the second-order central differentiation, which is generally less expensive than the higher-order scheme used in DNS, but also leads to wide use of finite volume methods that offer significant advantages in dealing with complex geometries. Using finite volume methods, computational techniques for complex geometries such as unstructured mesh can be conveniently implemented. Due to the flexibility offered by the numerical methods used, LES has been successfully applied to many industrial problems with complex geometries. In the following sections, two numerical techniques used in LES of incompressible flow in complex geometries are briefly discussed. The first technique was developed by Mahesh et al. (2004), using a finite volume method on unstructured grids, where the robustness of the method on skewed elements was particularly addressed. The second example concerns an immersed boundary technique, which is often used in conjunction with both the finite difference method and finite volume method for the simulation of flow interacting with solid boundaries.

A. LES on Unstructured Grids

In many CFD applications to complex geometries, an unstructured mesh is preferred. An unstructured mesh is a tessellation of a part of the Euclidean plane or Euclidean space by simple shapes, such as triangles or tetrahedra, in an irregular pattern, which can follow the shape of a body with complex geometry much more easily than a structured rectangular mesh. Mesh or grid of unstructured type can be used in CFD when the computational domain to be analyzed has an irregular shape, which is often encountered in practical applications. Mahesh et al. (2004) described a numerical method for LES of incompressible flows in complex geometries using a finite volume

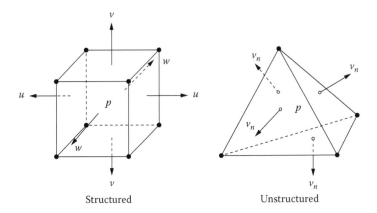

Structured   Unstructured

FIGURE 6.13   The staggered positioning of variables on an unstructured mesh of tetrahedra. (Mahesh et al. 2004; with permission from Elsevier Science Ltd.)

method based on an unstructured mesh, which was subsequently used in the LES of a gas-turbine combustor performed by Mahesh et al. (2006).

Unlike structured grids such as those used in finite difference schemes, unstructured grids require a list of the connectivity that specifies the order in which a given set of vertices make up individual elements. Figure 6.13 shows the staggered positioning of flow variables on an unstructured mesh of tetrahedra (Mahesh et al. 2004). In a tetrahedral element, there are four nodes, four faces, and five edges. The pressure and any scalars are stored at the circumcenter of the tetrahedron. The velocity component normal to each face v_n is stored at the circumcenter of each face. The correspondence to the classical staggered positioning of variables on structured grids is apparent. As pointed out by Mahesh et al. (2004), the convection term can be computed in velocity-vorticity (rotational) form on an unstructured mesh with staggered flow variables. However, the formulation has a few limitations. It is restrictive in that pressure or any other scalar is stored at the circumcenter of the triangular elements. This restricts the grid to elements whose circumcenter lies within them. Highly skewed elements can also cause problems since the circumcenter lies outside the element. Although projection of the velocity field is still possible in this situation, the inaccurate computation of the pressure gradient is not favorable. In addition, skewed elements render computations of the vorticity inaccurate. Also, the algorithm is restricted to tetrahedra. Although tetrahedra are very suitable for mesh with very complex geometries, Mahesh et al.

(2004) suggested that hexahedral elements are preferable due to the fact that hexahedral elements are more easily aligned with flow gradients such as boundary layers and it takes fewer hexahedral elements to fill space than comparable tetrahedra. In order to address the problems of the staggered rotational formulation, Mahesh et al. (2004) developed an alternative approach that can be implemented for hybrid grids of tetrahedra, hexahedra, wedges, and prisms. The basic idea was that the robustness of the method is determined essentially by the numerical discretization of the convection term, while robustness on skewed grids is determined by both discretization of the convection and the pressure-gradient terms. Discrete energy conservation (Mahesh et al. 2004) ensures that the flux of kinetic energy has contributions only from the boundary elements, which makes the solution robust without the use of numerical dissipation. For incompressible flows, discrete energy conservation refers to the fact that the convective and pressure terms in the discrete kinetic energy equation are expressible in divergence form, which is preferred for an incompressible solver. As stated by Mahesh et al. (2004), if the computational grid is sufficiently fine such as that used in a DNS where viscous dissipation is resolved on the grid, then discrete energy conservation is not essential. However, if the grid is not fine enough to resolve viscous dissipation such as that used in an LES, then discrete energy conservation is essential to obtain stable, accurate solutions.

In Mahesh et al. (2004), the Cartesian velocity components and pressure are stored at the centroids of the cells, while the face-normal velocities are treated as independent variables that are stored at the centroids of the faces. The methods can be outlined as follows. For a passive scalar ϕ, the spatial discretization of the convective term can be illustrated by the following equation:

$$\frac{\partial \phi}{\partial t} + \frac{\partial}{\partial x_i}(\phi u_i) = 0 \tag{6.6}$$

Using $\partial u_i / \partial x_i = 0$ for an incompressible flow and multiplying Equation (6.6) by ϕ yields

$$\frac{\partial \phi^2}{\partial t} + \frac{\partial}{\partial x_i}(\phi^2 u_i) = 0 \tag{6.7}$$

Equations (6.6) and (6.7) mean that conservation of ϕ implies conservation of ϕ^2. However, discretely, conserving ϕ does not automatically imply conservation of ϕ_2. Integrating Equation (6.6) over a cell and using the divergence theorem yields

$$V_{CV}\frac{d\phi_{CV}}{dt}+\sum_{faces\ of\ CV}\phi_{face}v_nA_f=0 \tag{6.8}$$

In Equation (6.8), ϕ_{CV} and V_{CV} denote the value of the scalar and the volume of a control volume CV, A_f is the face area, and v_n denotes the face-normal velocity in the direction of the outward normal at each face. Note that the incompressibility condition requires that $\sum_{faces}v_nA_f=0$. Also, ϕ is discretely conserved regardless of how ϕ_{face} is computed. However, ϕ^2 is discretely conserved only if the values of ϕ at the faces are calculated as a simple arithmetic mean of the values at the two cells CV and NB that have that particular face in common (Mahesh et al. 2004); that is,

$$\phi_{face}=\frac{\phi_{CV}+\phi_{NB}}{2} \tag{6.9}$$

The discrete equation for ϕ^2 is obtained by multiplying Equation (6.8) with ϕ_{CV} to obtain

$$V_{CV}\phi_{CV}\frac{d\phi_{CV}}{dt}+\frac{\phi_{CV}}{2}\sum_{faces\ of\ CV}(\phi_{CV}+\phi_{NB})v_nA_f=0 \tag{6.10}$$

Using $\sum_{faces}v_nA_f=0$, Equation (6.10) can be rewritten as

$$V_{CV}\frac{d\phi_{CV}^2}{dt}+\sum_{faces\ of\ CV}\phi_{CV}\phi_{NB}v_nA_f=0 \tag{6.11}$$

Summation of Equation (6.11) over all the cells in the computational domain yields

$$\sum_{volumes}\left(V_{CV}\frac{d\phi_{CV}^2}{dt}+\sum_{faces\ of\ CV}\phi_{CV}\phi_{NB}v_nA_f\right)=0 \tag{6.12}$$

The contribution from the interior faces cancels out in the second term to yield

$$\sum_{volumes} V_{CV} \frac{d\phi_{CV}^2}{dt} + \sum_{boundary\ faces} \phi_{CV}\phi_{NB}v_n A_f = 0 \tag{6.13}$$

In Equation (6.13), if the boundary conditions for the scalar ϕ are specified on the boundary faces, ϕ_{NB} can be defined as $\phi_{NB} = 2\phi_{face} - \phi_{CV}$.

As discussed by Mahesh et al. (2004), the interpolation $\phi_{face} = (\phi_{CV} + \phi_{NB})/2$ in Equation (6.9) is second-order accurate on uniform grids. On nonuniform grids, the interpolation could be weighted by the distances between the faces and the neighboring volumes. However, such weighted interpolation compromises the discrete energy conservation, which may lead to unstable solutions. On the contrary, the symmetric interpolation is both energy conserving and stable for arbitrarily nonuniform meshes, which are properties of prime importance for obtaining meaningful solutions in very complex geometries where mesh irregularities cannot always be avoided. In deriving Equation (6.11), $\sum_{faces} v_n A_f = 0$ was utilized since an incompressible flow was considered. The Poisson equation for pressure enforces this incompressibility constraint. If the Poisson equation is solved using direct methods, then the discrete divergence will be zero to machine accuracy. However, it is common to solve the Poisson equation iteratively; the discrete divergence in each computational cell is therefore determined by the tolerance to which the Poisson equation is converged. This has implications for discrete energy conservation. In this case, Equation (6.13) may be written as

$$\sum_{volumes} V_{CV} \frac{d\phi_{CV}^2}{dt} + \sum_{boundary\ faces} \phi_{CV}\phi_{NB}v_n A_f = -\sum_{volumes} \phi_{CV}^2 \sum_{faces\ of\ CV} v_n A_f \tag{6.14}$$

Note that even if the discrete divergence in each cell may be small, the collective contribution when summed over all the volumes in the computational domain can be significant. To avoid this cumulative effect, $\sum_{faces} v_n A_f = 0$ can be used when ϕ_{CV} is being advanced; that is, instead of using Equations (6.8) and (6.9) to advance ϕ_{CV}, the following equation can be used:

$$V_{CV}\frac{d\phi_{CV}}{dt}+\sum_{faces\ of\ CV}\frac{\phi_{NB}}{2}v_{n}A_{f}=0 \tag{6.15}$$

The above approach can be extended to the Navier–Stokes equations by computing the convection term in a similar manner. In particular, for $\phi=u_i$, this result indicates that there is no production of kinetic energy in the computational domain due to the numerics in the absence of time-discretization errors. A predictor–corrector formulation was derived by Mahesh et al. (2004) that emphasizes energy conservation for the convection and pressure terms on arbitrary grids. Accordingly, the cell velocities u_i and the face-normal velocities v_n defined at the center of the face are treated as essentially independent variables. Mahesh et al. (2004) used a predictor–corrector formulation along with explicit time-advancement given as

$$\frac{\hat{u}_{i}-u_{i}^{k}}{\Delta t}=\frac{1}{2}[3(NL+VIS)^{k}-(NL+VIS)^{k-1}] \tag{6.16}$$

In Equation (6.16), NL and VIS denote the nonlinear and viscous terms, and the superscript k represents the time step to be advanced, respectively. The predicted values of u_i are used to obtain predicted values for the face-normal velocities:

$$\hat{v}=\left(\frac{\hat{u}_{i}^{CV1}+\hat{u}_{i}^{CV2}}{2}\right)n_{i} \tag{6.17}$$

In Equation (6.17), $\hat{v}$ points from the volume $CV1$ to $CV2$. The predicted face-normal velocities are then projected using

$$\frac{v_{n}-\hat{v}}{\Delta t}=-\frac{\partial p}{\partial n} \tag{6.18}$$

The divergence-free constraint requires that $\sum_{faces}v_{n}A_{f}=0$. From Equation (6.18), the Poisson equation for pressure in integral form can be given as

$$\Delta t\sum_{faces\ of\ CV}\frac{\partial p}{\partial n}A_{f}=\sum_{faces\ of\ CV}\hat{v}A_{f} \tag{6.19}$$

The pressure Poisson equation given in Equation (6.19) can be solved iteratively. Once p is obtained, the Cartesian velocities are updated as

$$\frac{u_i^{k+1} - \hat{u}_i}{\Delta t} = -\frac{\partial p}{\partial x_i} \tag{6.20}$$

As pointed out by Mahesh et al. (2004), the method of computing $\partial p / \partial x_i$ affects the robustness of the solution on highly skewed grids. The computation of pressure always plays a crucial role in an incompressible flow solver. In complex geometry problems such as a gas-turbine combustor, the presence of skewed elements is inevitable. An obvious approach to computing the gradient at cell centers is to use the gradient theorem

$$\frac{\partial p}{\partial x_i} = \frac{1}{V} \sum_{faces} p_{face} A_f n_i \tag{6.21}$$

However, Equation (6.21) may lead to unstable solutions when applied on the highly skewed grids. This lack of robustness was understood by Mahesh et al. (2004) as from the contribution of the pressure gradient to the discrete kinetic energy equation. The contribution of the pressure gradient to discrete kinetic energy is not conservative in a nonstaggered grid formulation. Mahesh et al. (2004) derived a procedure for the evaluation of the pressure gradient in advancing the Cartesian velocities at the centers of the volumes. The approach to make the pressure-gradient term as energy conserving as possible is to satisfy this relation in a least-squares sense, that is, by minimizing

$$\sum_{faces\ of\ CV} \left(\left. \frac{\partial p}{\partial x_i} \right|_{iCV} n_i^{face} - \left. \frac{\partial p}{\partial n} \right|_{face} \right) A_f \tag{6.22}$$

This minimization allows $\partial p / \partial x_i$ to be computed in terms of the nearest neighbors and to be a local operation. Mahesh et al. (2004) found that the least-squares formulation was imperative to obtain robust, accurate solutions; unstable solutions were obtained in its absence.

B. LES with the Immersed Boundary Technique

Complex geometry treatment is at the center of the applications of LES to engineering problems. In the finite volume method by Mahesh et al. (2004), unstructured grids are used to approximate the complex flow boundaries where the grid lines or element surfaces are made to be aligned with the body surface, or as close as possible. There is also another numerical technique frequently used for complex geometries—the immersed boundary (IB) technique, as described by Moin (2002) in the context of LES for complex flows. The IB technique allows the computation of flow around complex objects without requiring the grid lines to be aligned with the body surface. The governing equations are solved on an underlying grid, which in principle can be structured or unstructured and which covers the entire computational domain without the bodies; no-slip boundary conditions are enforced via source terms or imaginary body forces in the equations, as described by Verzicco et al. (2000).

The IB technique can be incorporated into structured LES codes written in cylindrical, Cartesian, and curvilinear coordinates with and without zonal or mesh adaptation capability. In principle, it can also be incorporated into unstructured grids used in finite volume methods. The basic idea of the IB technique is given as follows.

In the immersed boundary technique, a boundary body-force term $\mathbf{f}$ is added to the incompressible equations to yield

$$\frac{D\bar{\mathbf{u}}}{Dt} = -\frac{1}{\rho}\nabla \bar{p} + \nabla \cdot \{\tilde{\nu}[\nabla \bar{\mathbf{u}} + (\nabla \bar{\mathbf{u}})^T]\} + \mathbf{f} \tag{6.23}$$

Equation (6.23) is simply the filtered Navier–Stokes momentum equations, equivalent to Equations (1.18)–(1.20) given in Chapter 1. In Equation (6.23), the effective viscosity $\tilde{\nu}$ is the sum of the molecular viscosity and the subgrid scale viscosity.

The time-discretized version of Equation (6.23) can be written as

$$\bar{\mathbf{u}}^{n+1} - \bar{\mathbf{u}}^n = \Delta t(\mathbf{RHS} + \mathbf{f}) \tag{6.24}$$

In Equation (6.24), Δt represents the computational time step, while the right-hand-side RHS contains the pressure and the nonlinear viscous terms, and the superscript denotes the time-step level. In the IB technique, in the region where we wish to mimic the solid body, in order to impose $\bar{\mathbf{u}}^{n+1} = \bar{\mathbf{v}}_b$ on the body with $\bar{\mathbf{v}}_b$ representing the velocity of the body, the forcing term $\mathbf{f}$ must satisfy

$$\mathbf{f} = -\mathbf{RHS} + \frac{\bar{\mathbf{v}}_b - \bar{\mathbf{u}}^n}{\Delta t} \tag{6.25}$$

In general, the surface of the region where $\overline{\mathbf{u}}^{n+1} = \overline{\mathbf{v}}_b$ does not coincide with a grid line or surface. The value of $\mathbf{f}$ at the node closest to the surface but outside the solid body is linearly interpolated between the value that yields $\overline{\mathbf{v}}_b$ on the solid body and zero in the interior of the flow domain where $\mathbf{f}$ has to vanish. This interpolation procedure is consistent with a centered second-order finite difference approximation, and the overall accuracy of the scheme remains second order.

To facilitate the application of the IB technique to complex configurations, a geometry preprocessor may be adopted in the solver. As a first step, the computational cells can be separated into "dead" (inside the body), "alive" (outside the body), and interface (partially inside). An automatic grid-refinement procedure can also be developed to improve the representation of the body on the underlying grid, as described by Moin (2002). In order to better represent the complex geometry, mesh adaptation capability may be implemented in the basic underlying CFD code. Presently, most LES codes do not have such a capability. Francois et al. (2004) used a multigrid technique with the IB technique in their computations of multiphase flows, which involves coupled momentum, mass, and energy transfer between moving and irregularly shaped boundaries, large property jumps between materials, and computational stiffness. Their immersed boundary technique is a combined Eulerian–Lagrangian method that was used to investigate the performance improvement by using the multigrid technique in the context of the projection method. They found that the multigrid technique speeds up the computation and furthermore the impact of the density ratio on the CPU time required is substantially reduced, but the impact of the viscosity ratio does not play a major role in the convergence rates.

There are recent developments in the IB technique. An IB technique for the simulation of flow interacting with a solid boundary was recently presented by Su et al. (2007). The formulation employs a mixture of Eulerian and Lagrangian variables, where the solid boundary is represented by discrete Lagrangian markers embedding in and exerting forces on the Eulerian fluid domain. The interactions between the Lagrangian markers and the fluid variables are linked by a simple discretized delta function, as described by Su et al. (2007). The numerical integration is based on a second-order fractional step method under the spatial staggered grid framework. Based on the direct momentum forcing on the Eulerian grids, a new force formulation on the Lagrangian marker is proposed, which ensures the satisfaction of the no-slip boundary condition on the immersed boundary in the intermediate time step. This forcing procedure involves solving a banded linear system of equations whose unknowns consist of the boundary forces on the Lagrangian markers.

A review of the IB methods was provided by Mittal and Iaccarino (2005). The IB method is often implemented together with the ghost-cell method. The ghost-cell method is an IB method based on a finite difference or a finite volume discretization. It introduces the presence of the immersed boundary by locally adapting the numerical fluxes using various interpolation techniques. For this purpose, virtual or "ghost" cells with an interpolated flow state are introduced. The boundary condition on the IB is enforced through the use of "ghost" cells, which are defined as cells in the solid that have at least one neighbor in the fluid. Current research in IB methods is focused toward improving their accuracy and efficiency. Using adaptive grid refinement with IB methods is also promising, especially for high Reynolds number flows. However, using local refinement increases the complexity of the algorithm and also begins to blur the line between IB methods and unstructured grid methods. As commented by Mittal and Iaccarino (2005), IB methods will see increased applications in complex turbulent flows, fluid-structure interaction, and multimaterial and multiphysics simulations.

REFERENCES

Balaras, E., Benocci, C., and Piomelli, U. 1996. Two-layer approximate boundary conditions for large-eddy simulations. *AIAA Journal* 34: 1111–1119.

Boussinesq, J. 1877. Théorie de l'écoulement tourbillant. *Mem. Présentés par Divers Savants Acad. Sci. Inst. Fr.* 23: 46–50.

Chakravarthy, V.K. and Menon, S. 2001. Large-eddy simulation of turbulent premixed flames in the flamelet regime. *Combustion Science and Technology* 162: 175–222.

Dejoan A. and Leschziner M.A. 2004. Large eddy simulation of periodically perturbed separated flow over a backward-facing step. *International Journal of Heat and Fluid Flow* 25: 581–592.

di Mare, F., Jones, W.P., and Menzies, K.R. 2004. Large eddy simulation of a model gas turbine combustor. *Combustion and Flame* 137: 278–294.

Domaradzki, J.A. and Loh, K.-C. 1999. The subgrid-scale estimation model in the physical space representation. *Physics of Fluids* 11: 2330–2342.

Ducros, F., Nicoud, F., and Poinsot, T. 1998. Wall-adapting local eddy viscosity models for simulation in complex geometries. In *6th ICFD Conference on Numerical Methods for Fluid Dynamics*, ed. M.J. Baines, 293–299. Oxford: Oxford University Computing Laboratory.

El-Asrag, H. and Menon, S. 2007. Large eddy simulation of bluff-body stabilized swirling non-premixed flames. *Proceedings of the Combustion Institute* 31: 1747–1754.

Francois, M., Uzgoren, E., Jackson, J., and Shyy, W. 2004. Multigrid computations with the immersed boundary technique for multiphase flows. *International Journal of Numerical Methods for Heat & Fluid Flow* 14: 98–115.

Germano, M., Piomelli, U., Moin, P., and Cabot, W. H. 1991. A dynamic sub-grid scale eddy viscosity model. *Physics of Fluids A* 3: 1760–1765.

Geurts, B.J. 2004. *Elements of direct and large-eddy simulation*. Philadelphia PA: Edwards.

Ghosal, S., Lund, T., Moin P. and Akselvoll, K. 1995. A dynamic localization model for large eddy simulation of turbulent flows. *Journal of Fluid Mechanics* 286: 229–255.

Hughes, T.J.R., Mazzei, L., Oberai, A.A. and Wray, A.A. 2001. The multiscale formulation of large eddy simulation: Decay of homogeneous isotropic turbulence. *Physics of Fluids* 13: 505–512.

Lilly, D.K. 1992. A proposed modification of the Germano subgrid-scale closure method. *Physics of Fluids A* 4: 633–635.

Mahesh, K., Constantinescu, G., Apte, S., Iaccarino, G., Ham, F., and Moin, P. 2006. Large-eddy simulation of reacting turbulent flows in complex geometries. *ASME Journal of Applied Mechanics* 73: 374–381.

Mahesh, K., Constantinescu, G., and Moin, P. 2004. A numerical method for large eddy simulations in complex geometries. *Journal of Computational Physics* 197: 215–240.

Menon, S., McMurtry, P.A., and Kerstein, A.R. 1993. A linear eddy flamelet subgrid model for large-eddy simulations of turbulent premixed combustion. In *Large Eddy Simulations of Complex Engineering and Geophysical Flows*, ed. B. Galperin, 288–314. Cambridge, UK: Cambridge University Press.

Menon, S., Yeung P.K. and Kim, W.W. 1996. Effect of subgrid models on the computed interscale energy transfer in isotropic turbulence. *Computer and Fluids* 25: 165–180.

Mittal, R. and Iaccarino, G. 2005. Immersed boundary methods. *Annual Review of Fluid Mechanics* 37:239–261.

Moin, P. 2002. Advances in large eddy simulation methodology for complex flows. *International Journal of Heat and Fluid Flow* 23: 710–720.

Moin, P., Squires, K., Cabot, W., and Lee, S. 1991. A dynamic sub-grid scale model for compressible turbulence and scalar transport. *Physics of Fluids A* 3: 2746–2757.

Rhie, C.M. and Chow, W.L. 1983. Numerical study of the turbulent flow past an airfoil with trailing edge separation. *AIAA Journal* 31: 1525–1533.

Sankaran, V. and Menon, S. 2005. LES of scalar mixing in supersonic mixing layers. *Proceedings of the Combustion Institute* 30: 2835–2842.

Schumann, U. and Sweet, R.A. 1988. Fast Fourier transforms for direct solution of Poisson's equation with staggered boundary conditions. *Journal of Computational Physics* 75: 123–137.

Smagorinsky, J. 1963. General circulation experiments with the primitive equations. *Monthly Weather Review* 91: 99–164.

Su, S.-W., Lai, M.-C., and Lin, C.-A. 2007. An immersed boundary technique for simulating complex flows with rigid boundary. *Computers and Fluids* 36: 313–324.

Valentino, M., Jiang, X., and Zhao, H. 2007. A comparative RANS/LES study of transient gas jets and sprays under diesel conditions. *Atomization and Sprays* 17: 451–472.

Verzicco, R., Mohd-Yusof, J., Orlandi, P., and Haworth, D. 2000. Large eddy simulation in complex geometric configurations using boundary body forces. *AIAA Journal* 38: 427–433.

Wang, M. and Moin, P. 2002. Dynamic wall modelling for large eddy simulation of complex turbulent flows. *Physics of Fluids A* 14: 2043–2051.

CHAPTER 7

LES of Compressible Flows

As discussed in Chapter 5, compressible flows represent a broad range of fluid flows of relatively high speeds. Although compressibility in liquid fluids is often negligible, there is a situation where the pressure variation in the flow field is large enough to cause substantial changes in the density of the fluid. This is known as the cavitation phenomenon, where the pressure variations in the flow are large enough to cause a phase change. For liquids, whether the incompressible assumption is valid depends on the fluid properties, particularly the pressure and temperature of the fluid and how close they are to the critical pressure and temperature. However, the majority of the practical compressible flows are gas flows. For flow of gases, the compressibility needs to be taken into account at Mach numbers above approximately 0.3. Such compressible flows are of great importance to aerospace engineering and many other high-speed flow applications.

Many high-speed flows have Reynolds numbers too high for DNS to be a viable option, given the state of computational power for the next few decades. In many aeronautical applications, DNS remains impossible for practical applications. For instance, the complete aerodynamic computation of transport aircraft wings such as those on an Airbus A300 or Boeing 747 having nominal Reynolds numbers of 40 million (based on the wing chord) at flight conditions are well beyond the limit of DNS. Even for a small aircraft with a dimension greater than 3 meters, moving faster than 72 km/h or 20 m/s with a nominal Reynolds number of the flow above 4 million is out of reach for DNS. In order to solve the entire domain of these real-life flow problems or even if only a small part of the flow, RANS, LES, and detached eddy simulation (DES) as a combination of RANS and LES are a necessity for the foreseeable future. The main advantage of LES over computationally cheaper RANS approaches is the increased level of details it can deliver, especially the prediction of flow unsteadiness, which is quite important to aerodynamic applications. While RANS methods provide "averaged" results, LES is able to predict instantaneous flow characteristics and resolve turbulent flow structures. This is particularly important in simulations involving chemical reactions, such as the combustion of fuel in an internal combustion engine or an aircraft engine. While the "averaged" concentration of chemical species may be too low to trigger a reaction, instantaneously there can be localized areas of high concentration in which reactions can take place. For aeroacoustic predictions, LES also offers significantly more meaningful results over RANS since sound radiation is essentially the propagation of instantaneous pressure waves. Computational aeroacoustic problems require the formation allowing compressibility, since sound waves can be found only from the fluid equations of compressible flows.

As shown in Chapter 1, the governing equations for compressible flows are significantly different from those for incompressible flows. Compared with LES of incompressible flows (Sagaut 2006), LES of compressible flows is receiving increasing attention, but it is still relatively scarce. For compressible flows, the strong coupling between fluid density and other flow quantities is a feature that is absent for incompressible flows. A time- or ensemble-averaging or spatial filtering of the compressible flow governing equations using directly the Reynolds decomposition of all the flow variables leads to a very complex format of the equations. However, this unnecessary complexity can be avoided by using Favre averaging or filtering. For the LES of compressible flows, there can be Favre-filtered Navier–Stokes equations. Using Favre filtering, the filtered governing equations of compressible flows are very similar to those of incompressible flows. Consequently, the subgrid scale (SGS) modeling of compressible flows becomes very similar to that of incompressible flows. For instance, the Boussinesq eddy viscosity hypothesis (Boussinesq 1877) for subgrid-scale turbulent stresses can be used, which relates the Reynolds stresses to the mean velocity gradients as shown in Equation (6.2) and the subgrid scale eddy viscosity can be modeled with an SGS model. There are strong links between the SGS modeling of incompressible and compressible flows. For instance, the dynamic approach developed by Germano et al. (1991) for incompressible flows was successfully extended to compressible flows by Moin et al. (1991).

This chapter is devoted to LES of compressible flows. Similar to Chapter 6, some sample LES results for both reacting and nonreacting flows are presented first. The SGS modeling, which is the most distinctive feature of LES, is then discussed in the context of compressible flows. In the discussions on SGS modeling, an important concept in LES, implicit large-eddy simulation (ILES), is included in this chapter. ILES has provided many numerical simulations with an efficient and effective model for turbulence. For compressible flows, the capacity for ILES has been shown to arise from a broad class of numerical methods with specific properties producing nonoscillatory solutions using limiters that provide these methods with nonlinear stability. However, it is worth noting that ILES is not restricted to compressible flows. As discussed by Rider (2007), much of the understanding of ILES modeling has proceeded in the realm of incompressible flows. The ILES of compressible flows can be considered as dominated by an effective self-similarity subgrid model, like the incompressible flow. The efficacy of ILES as a model for compressible flows was analyzed by Rider (2007), where the model can have several limits including the incompressible limit. Finally, it is worth mentioning that discussions on numerical features of the spatial discretization and temporal integration schemes are omitted in this chapter, because these were covered in previous chapters.

I. SAMPLE RESULTS OF LES OF COMPRESSIBLE FLOWS

A. LES of a Ramjet Combustor (Roux et al. 2008)

A ramjet combustor is an essential part of a ramjet engine. As the power plant for a high-supersonic vehicle, a ramjet engine consists of three major components: inlet, combustor, and nozzle. Unlike a turbofan or turbojet engine, the ramjet does not have the compressor or turbine. Air enters the inlet, where it is compressed, and then enters the combustion zone, where it is mixed with the fuel and burnt. The hot gases are then expelled through the nozzle, producing thrust. The operation of the ramjet depends on the inlet to decelerate the incoming air to raise the pressure in the combustion zone. The higher the velocity of the incoming air, the greater the pressure rise is. This is why the ramjet operates best at high supersonic velocities. At subsonic velocities, the ramjet is inefficient. The combustion process in an ordinary ramjet takes place at low supersonic velocity. At high supersonic velocities, a very large pressure rise is developed that is more than sufficient to support operation of the ramjet. Since the inlet has to decelerate high supersonic velocities to low subsonic velocities, large pressure losses could occur. The deceleration process also produces a temperature rise. At some limiting speed, the temperature will approach the limit set by the wall materials and cooling methods. In the past few years, research and development have been done on a ramjet that has the combustion process taking place at supersonic velocities using a supersonic combustion process. The objective is to reduce the deceleration and the associated pressure loss in the inlet and also the temperature rise. This is known as scramjet. It should be noted that since large velocities are required to start ramjets, another engine system is required to accelerate an aircraft propelled by a ramjet.

Roux et al. (2008) performed LES of the flow in a side-dump ramjet combustor, which is a two-inlet configuration burning gaseous propane with air. Three computational cases were performed, including reacting and nonreacting cases. The combustor configuration is shown in Figure 7.1.

The time-dependent compressible Navier–Stokes equations are filtered using a Favre filtering operation yielding the LES equations, which include chemical reactions. The governing equations include subgrid-scale quantities that need to be modeled. The unresolved SGS stresses are modeled using the Boussinesq eddy viscosity assumption (Boussinesq 1877) as given in Equation (6.2). The wall-adapting local eddy-visosity (WALE) model of Ducros et al. (1998) and Nicoud and Ducros (1999) was chosen to model the SGS viscosity. The parallel LES simulation performed by Roux et al. (2008) solves the governing equations using a cell-vertex finite volume approximation. The numerical integration uses Lax–Wendroff type or Taylor–Galerkin weighted residual central distribution schemes. The Taylor–Galerkin scheme, as described by Colin and Rudgyard (2000), provides third-order accuracy on hybrid meshes and is particularly adequate for low-dissipation requirements of LES applications. Time integration is done by a third-order explicit multistage Rung–Kutta scheme. Since the ramjet flow contains a choked nozzle, the outlet flow is supersonic and no shock capture method is needed.

The mesh used for the numerical simulations contains 1.2 million tetrahedra, as shown in Figure 7.2, with local refinement to ensure grid resolution in accordance with the requirements for the LES. The nozzle walls are handled as slip walls. All other walls correspond to no-slip adiabatic surfaces. At the inlets, the Navier–Stokes characteristic boundary conditions (NSCBC) are used to ensure a physical representation of the acoustic wave propagation. The outlet nozzle is included in the LES domain. This

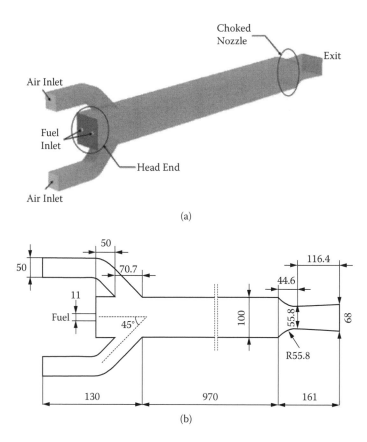

(a)

(b)

FIGURE 7.1 Schematic of the experimental ramjet combustor along with its dimensions. (Roux et al. 2008; with permission from Elsevier Science Ltd.)

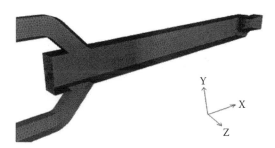

FIGURE 7.2 Three-dimensional view of the ramjet combustor mesh. (Roux et al. 2008; with permission from Elsevier Science Ltd.)

FIGURE 7.3 Isosurfaces of axial velocity for the nonreacting case of the ramjet combustor. (Roux et al. 2008; with permission from Elsevier Science Ltd.)

avoids all uncertainties on the acoustic behavior of the outlet boundary, which is supersonic and thus well posed mathematically and numerically. This is not the case for the air inlet boundary conditions, which must be nonreflective to prevent artificial forcing of the acoustic field within the domain of computation.

In Figure 7.3, the isosurfaces of axial velocity are shown for the nonreacting case. The obtained envelope encompasses the two high-velocity streams coming from the air inlet ducts, which impact each other within the chamber before reaching the two side walls of the ramjet. This isosurface is characteristic of crushing jets coalescing into a jet sheet. In this specific case, the generated high-velocity sheet impacts the vertical walls of the combustion chamber. The main recirculation zone is localized near the head-end of the combustor and is evidenced by a low value of the velocity magnitude isosurface (in light gray). Other recirculating regions are created just downstream of the air inlets, on the top and bottom walls of the chamber. Two recirculating bubbles also appear within the two airstream ducts. These flow patterns are strongly linked to the combustor geometry. Indeed, the sudden inclinations of the duct air inlets induce detachment points of the injected air flow. Behind these points, small recirculation zones diminish the effective flow passage, hence increasing the air velocity before the air enters the combustion chamber. Similarly, the sudden expansion seen by the incoming flow of air when entering the main chamber duct explains the two recirculation bubbles appearing on the top and bottom floors of the combustor.

For this cold flow (nonreacting case) configuration, the unsteady turbulent behavior is dominated by the oscillation of the impinging jets.

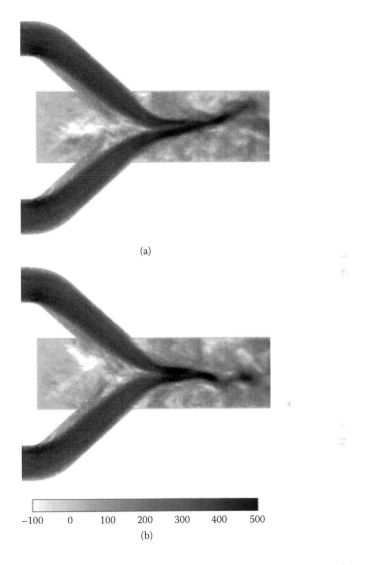

(a)

(b)

−100   0   100   200   300   400   500

FIGURE 7.4   Instantaneous axial velocity for the nonreacting case in the symmetry plane of the ramjet combustor. (Roux et al. 2008; with permission from Elsevier Science Ltd.)

This highly unsteady feature is illustrated in Figure 7.4, where two snapshots are taken at instants when the coalesced jet is deviated toward the top and bottom walls of the main chamber (top and bottom subfigures, respectively). The actual characterization of such a phenomenon remains

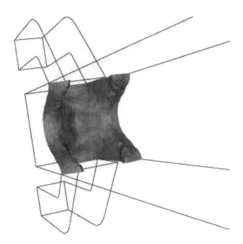

FIGURE 7.5 Isosurface of the averaged fuel mass fraction for the reacting case of the ramjet combustor. (Roux et al. 2008; with permission from Elsevier Science Ltd.)

difficult, since acoustic and aerodynamic phenomena have similar time scales.

From the LES results, mean flow quantities can be conveniently obtained. Figure 7.5 shows an isosurface for a mean-averaged fuel mass fraction at 0.15. The pattern of the isosurface evidences the path traveled by the fuel from the head-end to the chamber. Hot burned gas pockets and combustion products are convected toward the chamber exhaust.

B. LES of Compressible Flows around Cavities (Rubio et al. 2006)

Cavity flows are encountered in a broad range of applications. The phenomenon of flows passing over a cavity occurs in applications, including transport systems, aircraft wheel systems, and other aerospace applications. Rubio et al. (2006) performed a compressible LES of a flow passing over a cavity for different cavity configurations. The compressible viscous Navier–Stokes equations are solved. The governing equations include a Favre-filtered part and an unresolved part that is modeled with a subgrid-scale model. A Smagorinsky (1963) model with constant coefficients is employed while the filter size is set equal to the mesh size.

The governing equations are integrated in time using a fourth-order explicit Runge–Kutta scheme. Convective and viscous terms are discretized using second-order schemes. On the walls, isothermal, nonslip,

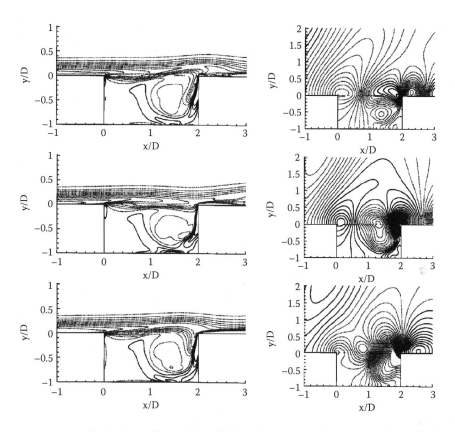

FIGURE 7.6 Evolution of vorticity (left) and pressure fluctuation (right) of a rectangular cavity; progressive time instants are shown from top to bottom. (Rubio et al. 2006; with permission from John Wiley & Sons, Ltd.)

no-penetration conditions are used. At the inlet, the characteristic soft boundary conditions are used while at the outflow, the subsonic characteristic outflow condition of the NSCBC type is used. A sponge zone is also used to eliminate the spurious wave reflections. At the far-side boundary, a free boundary condition is applied where the derivatives of the primitive variables are set to zero.

Figure 7.6 shows the instantaneous vorticity and pressure contours at three different time instants. The vorticity contours show a steady vortex or a recirculation zone occupying the rear half of the cavity, indicating that the interaction of the shear layer with the flow inside the cavity is very weak. The pressure contours allow the identification of what will be the directivity of the acoustic field propagated upstream far from the cavity. The regions in the cavity with negative values of the pressure fluctuations

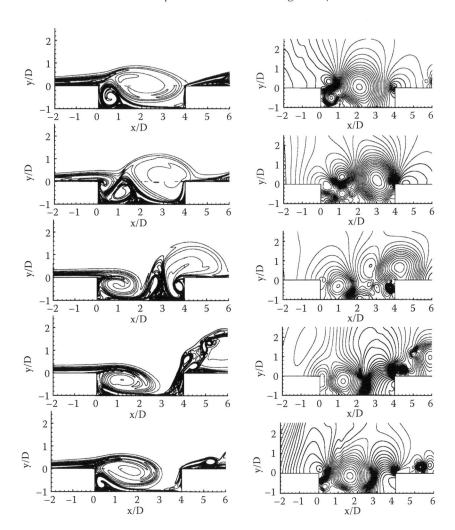

FIGURE 7.7 Evolution of vorticity (left) and pressure fluctuation (right) of a lengthened rectangular cavity; progressive time instants are shown from top to bottom. (Rubio et al. 2006; with permission from John Wiley & Sons, Ltd.)

are the result of the swirling movement, leading to the formation of the stationary recirculation zone.

Figure 7.7 shows the instantaneous vorticity and pressure contours at five different time instants for a lengthened cavity (compared to Figure 7.6). The flow is characterized by a large-scale vortex shedding from the cavity leading edge. The vortex reaches nearly the cavity size, dragging during its formation irrotational free-stream fluid into the cavity. The vortex is then

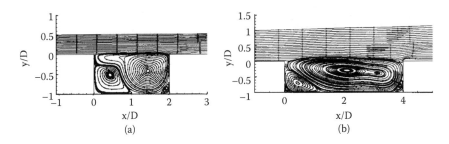

FIGURE 7.8 Time-averaged streamlines for both cavity configurations. (Rubio et al. 2006; with permission from John Wiley & Sons, Ltd.)

shed from the leading edge and violently ejected from the cavity. The time-averaged streamlines for both cavity configurations are shown in Figure 7.8. For the small-length cavity shown in Figure 7.8(a), the mean flow streamlines are almost horizontal along the mouth of the cavity, while they are deflected in the lengthened cavity case as shown in Figure 7.8(b).

II. SUBGRID-SCALE MODELING OF COMPRESSIBLE FLOWS AND IMPLICIT LARGE-EDDY SIMULATION (ILES)

A. Subgrid-Scale Modeling of Compressible Flows

For compressible and/or reacting flows, density is an important variable and the density weighted averaging or density weighted filtering plays a significant role in RANS turbulence modeling and in LES subgrid-scale turbulence modeling. In density weighted Favre averaging, a dependent variable f can be decomposed into a mean part $\tilde{f}$ and a fluctuating part f'' using a density weighted average as f $f = \tilde{f} + f''$, $\tilde{f} = \overline{\rho f}/\overline{\rho}$, where the overbars denote averages using the Reynolds decomposition and the auxiliary relations include $\overline{\rho f''} = 0$ and $\overline{\rho \tilde{f}} = \overline{\rho}\tilde{f} = \overline{\rho f}$. Note that Favre averaging or filtering does not apply to density itself and the pressure by convention. Similar to the Favre averaging in the RANS approach, LES for compressible and/or reacting flows involves the spatial Favre filtering defined as

$$\tilde{f}_i = \frac{1}{\overline{\rho}_i} \int \rho_i(x') G(x, x') f_i(x') dx' \tag{7.1}$$

where G denotes the filter function. Applying the spatial Favre filtering to the governing equations for an unsteady three-dimensional compressible and viscous flow without body forces, the following equations can be obtained:

The continuity equation,

$$\frac{\partial \bar{\rho}}{\partial t} + \frac{\partial \bar{\rho} \tilde{u}_i}{\partial x_i} = 0 \tag{7.2}$$

Momentum equations (the Navier–Stokes equations),

$$\frac{\partial (\bar{\rho} \tilde{u}_i)}{\partial t} + \frac{\partial (\bar{\rho} \tilde{u}_i \tilde{u}_j)}{\partial x_j} = -\frac{\partial \bar{p}}{\partial x_i} + \frac{\partial (\tilde{\tau}_{ij} + \tau_{ij,SGS})}{\partial x_j} \tag{7.3}$$

The energy equation,

$$\frac{\partial (\bar{\rho} \tilde{E}_T)}{\partial t} + \frac{\partial (\bar{\rho} \tilde{E}_T \tilde{u}_i)}{\partial x_i} = -\frac{\partial (\bar{p} \tilde{u}_i)}{\partial x_i} + \frac{\partial [(\tilde{\tau}_{ij} + \tau_{ij,SGS}) \tilde{u}_i]}{\partial x_j} - \frac{\partial (\tilde{q}_i + q_{i,SGS})}{\partial x_i} \tag{7.4}$$

In Equation (7.4), the Favre filtered total energy per mass unit $\tilde{E}_T$ is defined as $\tilde{E}_T = \bar{p}/[(\gamma - 1)\bar{\rho}] + (\tilde{u}_i \tilde{u}_i)/2$.

In the mathematical description of compressible turbulent flows, the application of the filtering operation to the instantaneous set of compressible Navier–Stokes transport equations yields the LES transport equations, which contain the so-called SGS quantities, $\tau_{ij,SGS}$ and $q_{i,SGS}$, that need modeling (Saqaut, 2006). The unresolved SGS stress tensors $\tau_{ij,SGS}$ can be modeled using the Boussinesq assumption (Boussinesq 1877; Pope 2000; Smagorinsky 1963), given by

$$\tau_{ij,SGS} - \frac{1}{3}\tau_{kk,SGS}\delta_{ij} = -\bar{\rho} v_T \left(\frac{\partial \tilde{u}_i}{\partial x_j} + \frac{\partial \tilde{u}_j}{\partial x_i} \right) = -2\bar{\rho} v_T \tilde{S}_{ij} \tag{7.5}$$

The SGS heat flux $q_{i,SGS}$ can be modeled as

$$q_{i,SGS} = -\frac{\mu_t c_p}{\Pr_t} \frac{\partial \tilde{T}}{\partial x_i} \tag{7.6}$$

In Equation (7.6), $\Pr_t$ represents the turbulent Prandtl number. To determine the turbulent viscosity, a Smagorinsky model can be used, where

$$v_T = (C_s \Delta)^2 |\tilde{S}| = (C_s \Delta)^2 \sqrt{2\tilde{S}_{ij}\tilde{S}_{ij}} \tag{7.7}$$

In Equation (7.7), C_S is the Smagorinsky constant and Δ is the filter width, as they were in Chapter 6.

From the above analysis, it is clear that when Favre filtering is used, the SGS turbulence modeling of compressible flows is very similar to that of incompressible flows. Dynamic SGS models can also be proposed for LES of compressible flows. For example, the dynamic SGS model of Germano et al. (1991) was generalized by Moin et al. (1991) for LES of compressible flows and the transport of a scalar. The model was applied to LES of decaying isotropic turbulence, and the results are in excellent agreement with experimental data and DNS results.

The subgrid scale models for compressible flows were often developed based on SGS models for incompressible flows. Erlebacher et al. (1992) developed subgrid-scale models for LES of compressible turbulent flows based on the Favre filtered equations of motion for an ideal gas. A compressible generalization of the linear combination of the Smagorinsky model and scale similarity model, in terms of Favre filtered fields, was obtained for the subgrid scale stress tensor (Erlebacher et al. 1992). An analogous thermal linear combination model was also developed for the subgrid-scale heat flux vector. The two dimensionless constants associated with these SGS models are obtained by correlating with the DNS results of compressible isotropic turbulence.

B. Implicit Large-Eddy Simulation (ILES)

Implicit large-eddy simulation (ILES) is a relatively new approach to LES; it is conceptually different from LES where an explicit SGS model is employed. In ILES, the truncation error of the discretization of the convective terms functions as a subgrid-scale model. Therefore, the model is implicitly contained within the discretization, and an explicit computation of model terms becomes unnecessary. ILES combines generality and computational efficiency with documented success in many areas of complex fluid flow. In general, the SGS model in an LES operates on a range of scales, which is marginally resolved by discretization schemes. Accordingly, the discretization scheme and the subgrid-scale model are linked. One can exploit this link by developing discretization methods from subgrid-scale models, or the converse. Approaches where SGS models and numerical discretization are fully merged are called implicit LES.

Although SGS modeling is the most distinctive feature of an explicit LES, an SGS model is not needed in ILES. Hahn and Drikakis (2004) presented a numerical investigation of high-resolution schemes for solving the compressible Euler and Navier–Stokes equations in the context of ILES, or monotonically integrated LES (MILES) in the context of compressible flows. Hahn and Drikakis (2004) discussed three high-resolution schemes: a flux vector splitting (FVS), a characteristics-based (Godunov-type), and a hybrid total variation diminishing (TVD) scheme.

In order to explain the ILES method, the fundamental governing equations for an unsteady, three-dimensional, compressible, and viscous flow can be put into a generic vector form:

$$\frac{\partial \mathbf{U}}{\partial t} + \frac{\partial \mathbf{E}}{\partial x} + \frac{\partial \mathbf{F}}{\partial y} + \frac{\partial \mathbf{G}}{\partial z} = \mathbf{H} \tag{7.8}$$

where the vectors **U**, **E**, **F**, **G**, and **H** are defined as

$$\mathbf{U} = \begin{pmatrix} \rho \\ \rho u \\ \rho v \\ \rho w \\ E_T \end{pmatrix} \tag{7.9}$$

$$\mathbf{E} = \begin{bmatrix} \rho u \\ \rho u^2 + p - \tau_{xx} \\ \rho u v - \tau_{xy} \\ \rho u w - \tau_{xz} \\ (E_T + p)u + q_x - u\tau_{xx} - v\tau_{xy} - w\tau_{xz} \end{bmatrix} \tag{7.10}$$

$$\mathbf{F} = \begin{bmatrix} \rho v \\ \rho u v - \tau_{xy} \\ \rho v^2 + p - \tau_{yy} \\ (E_T + p)v + q_y - u\tau_{xy} - v\tau_{yy} - w\tau_{yz} \end{bmatrix} \tag{7.11}$$

$$
\mathbf{G} = \begin{bmatrix}
\rho w \\
\rho u w - \tau_{xz} \\
\rho v w - \tau_{yz} \\
\rho w^2 + p - \tau_{zz} \\
(E_T + p)w + q_z - u\tau_{xz} - v\tau_{yz} - w\tau_{zz}
\end{bmatrix}
\tag{7.12}
$$

$$
\mathbf{H} = \begin{pmatrix}
0 \\
\rho f_x \\
\rho f_y \\
\rho f_z \\
\rho \mathbf{f} \cdot \mathbf{V}
\end{pmatrix}
\tag{7.13}
$$

For the numerical solution of Equation (7.8), an explicit third-order Runge–Kutta scheme (Shu and Osher 1988) can be used for the time integration, and central differences can be used for the viscous terms, while different high-resolution schemes for the discretization of the advective terms may be used as described by Hahn and Drikakis (2004). As an example, the numerical schemes for the discretization of the advective terms are briefly described below, but more details can be found in Bagabir and Drikakis (2004).

Considering the one-dimensional, inviscid counterpart of Equation (7.8), $\partial \mathbf{U}/\partial t + \partial \mathbf{E}/\partial x = 0$ where $\mathbf{U}$ is the array of the unknown variables and $\mathbf{E}$ is the flux associated with the terms in x direction. The advective flux derivative $\partial \mathbf{E}/\partial x$ (similarly, for the other advective flux derivatives $\partial \mathbf{F}/\partial y$ and $\partial \mathbf{G}/\partial z$) is discretized at the center of the ith control volume or the grid point using the values of the intercell fluxes, that is, $\partial \mathbf{E}/\partial x = (E_{i+1/2} - E_{i-1/2})/\Delta x$. The definition of the intercell flux function distinguishes among the different high-resolution schemes employed:

- Flux vector splitting (FVS) scheme: The flux vector splitting is a technique for achieving upwind bias in numerical flux function. The FVS scheme is built by adding the contributions of both cells located on either side of a given interface and it defines the intercell advective flux as $\mathbf{E}_{i+1/2}^{SW-FVS} = \mathbf{E}_{i+1/2}^{+}(\mathbf{U}_L) + \mathbf{E}_{i+1/2}^{-}(\mathbf{U}_R)$, where the superscript $SW-FVS$ represents the Steger-Warming FVS scheme (Bagabir and Drikakis 2004; Zóltak and Drikakis 1998), while the left and right conservative variables $\mathbf{U}_L$ and $\mathbf{U}_R$ can be obtained by accurate interpolation schemes.

- Characteristics-based scheme: The scheme is a Godunov-type method (Drikakis and Rider 2005) that defines the conservative variables along the characteristics as functions of their characteristic values. A high-order interpolation scheme such as the third-order interpolation scheme (Zóltak and Drikakis 1998) can be used to compute the characteristic values, depending on the sign of the characteristic speed or the eigenvalue.

- The hybrid TVD scheme: It defines the advective flux as $\mathbf{E}_{i+1/2} = \psi_{i+1/2}\mathbf{E}_{i+1/2}^{(SW-FVS)} + (1-\psi_{i+1/2})\mathbf{E}_{i+1/2}^{CB}$, where $\mathbf{E}_{i+1/2}^{(SW-FVS)}$ and $\mathbf{E}_{i+1/2}^{CB}$ are the intercell fluxes according to the SW-FVS and characteristics-based (CB) schemes. The term $\psi_{i+1/2}$ is a limiter function defined by the square of the local Mach number differences across cell faces (Zóltak and Drikakis 1998). Limiters are the general nonlinear mechanism that distinguishes modern methods from classical linear schemes. Their role is to act as a nonlinear switch between more than one underlying linear method, thus adapting the choice of numerical method based upon the behavior of the local solution. Limiters result in nonlinear methods even for linear equations in order to achieve second-order accuracy simultaneously with monotonicity. Numerical flux limiters can act like dynamic, self-adjusting models, modifying the numerical viscosity to produce a nonlinear eddy viscosity.

In an ILES, the numerical scheme has to be constructed such that the leading order truncation errors satisfy physically required SGS model properties, and hence nonlinear discretization procedures are required due to the nonlinear characteristics of the SGS Reynolds stresses. Finite volume versions of shock-capturing schemes designed under the requirements of convergence to weak solution while satisfying the entropy condition schemes can be viewed as relevant for ILES. In the monotonically integrated LES, or MILES, the effects of the SGS physics on the resolved scales are incorporated in the functional reconstruction of the convective fluxes using locally monotonic methods. Analysis based on the modified equations can be used to demonstrate an intriguing feature of MILES, namely that when based on a particular class of flux-limiting schemes, the convection discretization implicitly generates a nonlinear tensor-valued eddy viscosity that acts to stabilize the flow and suppress unphysical oscillations (Grinstein 2003). MILES may be extended to the more general concept of nonlinear ILES in which the functional reconstruction of the convective flux functions is carried out using high-resolution nonlinear numerical schemes incorporating a sharp velocity-gradient capturing capability operating at the smallest resolved scales. By focusing on the inviscid inertial-range dynamics and on regularization of the under-resolved flow, ILES follows up naturally on the historical precedent of using this kind of numerical scheme for shock capturing.

Grinstein (2003) also pointed out the challenges for ILES development, which include developing a common appropriate mathematical and physical framework for its analysis and development, further understanding the connections between implicit SGS model and numerical scheme and, in particular, addressing how to build physics into the numerical scheme to improve on global ILES performance, such as the implicitly implemented SGS dissipation and backscatter features. Recently, Hickel et al. (2008) proposed a systematic framework for the design, analysis, and optimization of nonlinear discretization schemes for implicit LES. In this framework, parameters inherent to the discretization scheme are determined in such a way that the numerical truncation error acts as a physically motivated SGS model. The resulting so-called adaptive local deconvolution method (ALDM) for implicit LES allows for reliable predictions of isotropic forced and decaying turbulence and of unbounded transitional flows for a wide range of Reynolds numbers (Hickel et al. 2008). Deconvolution parameters are determined by an analysis of the spectral numerical viscosity. An automatic optimization based on an evolutionary algorithm is employed to obtain a set of parameters that results in an optimum spectral match for the numerical viscosity with theoretical predictions for isotropic turbulence (Hickel et al. 2006). Although model parameters of ALDM have been determined for isotropic turbulence at infinite Reynolds number, it successfully predicts mean flow and turbulence statistics in the considered physically complex, anisotropic, and inhomogeneous flow regime.

It has been shown that the implicit model in ILES performs at least as well as an established explicit SGS model in a few turbulent flows (Grinstein et al. 2007), which may also be combined together. For instance, additional explicit SGS modeling might be needed to address inherently small-scale physical phenomena such as scalar mixing and combustion, which are actually outside the realm of any LES approach, leading to efficient "mixed" SGS models incorporating both explicit SGS modeling and implicit modeling through the numerics.

The development of an ILES approach calls for a better understanding of the behavior of the numerical schemes and their relationships to the filtering process in LES theory. The use of a higher-order scheme leads to a fewer number of grid points in order to achieve a required level of accuracy and hence a larger filter width accordingly. An early theory for the filtering processing based on the truncation errors and mesh sizes of high-order schemes still needs to be fostered. As a new approach to computing turbulent fluid dynamics, ILES combines generality and computational efficiency with documented success in many areas of complex fluid flows. Grinstein at al. (2007) edited a text on implicit large-eddy simulation, which synthesizes the current understanding of the theoretical basis of the ILES methodology and reviews its accomplishments. More details and applications of ILES can be found in Grinstein at al. (2007), which represents a comprehensive description of the state-of-the-art methodology.

Finally, it is worth noting that the approach of ILES also raises a question of how to define the numerical transition between DNS and LES. A typical definition for DNS is to resolve all the relevant time and length scales in the flow field, which requires the grid size to be some small multiples of the Kolmogorov scale (Drikakis and Rider, 2005). In both DNS and LES, high-accuracy fluid solvers based on centered, compact, or spectral schemes are employed, where numerical dissipation can be minimized. In DNS, physical viscosity ideally provides all the dissipation necessary to ensure numerical stability, which can normally be achieved when a very fine mesh is employed. For ILES, the governing equations solved are the same as those for DNS without the SGS modeling. However, there are a few differences. First, a very fine mesh is not needed in ILES since the small scales do not need to be resolved. Second, ILES does not need to employ very high-order numerical schemes because it is not required to resolve the small scales. As stated in Chapter 1, in state-of-the-art DNS, the discretization schemes used are at least fourth order, typically sixth and above. In LES, the numerical schemes used are normally of lower order. Third, truncation error of the discretization of the convective terms in ILES functions as an SGS model, which not only needs to be stable enough so that the energy in the smallest resolved scales will not grow unbounded leading to divergence, but also needs to be able to represent the subgrid-scale contribution.

REFERENCES

Bagabir, A. and Drikakis, D. 2004. Numerical experiments using high-resolution schemes for unsteady, inviscid, compressible flows. *Computer Methods in Applied Mechanics and Engineering* 193: 4675–4705.

Boussinesq, J. 1877. Théorie de l'écoulement tourbillant. *Mem. Présentés par Divers Savants Acad. Sci. Inst. Fr.* 23: 46–50.

Colin, O. and Rudgyard, M. 2000. Development of high-order Taylor–Galerkin schemes for LES. *Journal of Computational Physics* 162: 338–371.

Drikakis, D. and Rider, W. 2005. *High-resolution methods for incompressible and low-speed flows.* Berlin: Springer.

Ducros, F., Nicoud, F., and Poinsot, T. 1998. Wall-adapting local eddy viscosity models for simulation in complex geometries. In *6th ICFD Conference on Numerical Methods for Fluid Dynamics,* ed. M.J. Baines, 293–299. Oxford: Oxford University Computing Laboratory.

Erlebacher, G., Hussaini, M.Y., Speziale, C.G., and Zang, T.A. 1992. Towards the large-eddy simulation of compressible turbulent flows. *Journal of Fluid Mechanics* 238: 155–185.

Germano, M., Piomelli, U., Moin, P., and Cabot, W.H. 1991. A dynamic sub-grid scale eddy viscosity model. *Phys. Fluids A* 3: 1760–1765.

Grinstein, F.F. 2003. On implicit subgrid scale modeling for turbulent flows. Technical Report CaltechLESSGS: Session A.2. CA: California Institute of Technology.

Grinstein, F.F., Margolin, L.G., and Rider, W.J. (editors). 2007. *Implicit large eddy simulation.* Cambridge, UK: Cambridge University Press.

Hahn, M. and Drikakis, D. 2004. Large eddy simulation of compressible turbulence using high-resolution methods. *International Journal for Numerical Methods in Fluids* 47: 971–977.

Hickel, S., Adams, N.A., and Domaradzki, J.A. 2006. An adaptive local deconvolution method for implicit LES. *Journal of Computational Physics* 213: 413–436.

Hickel, S., Kempe, T., and Adams, N.A. 2008. Implicit large-eddy simulation applied to turbulent channel flow with periodic constrictions. *Theoretical and Computational Fluid Dynamics* 22: 227–242.

Moin, P., Squires, K., Cabot, W., and Lee, S. 1991. A dynamic sub-grid scale model for compressible turbulence and scalar transport. *Physics of Fluids A* 3: 2746–2757.

Nicoud, F. and Ducros, F. 1999. Subgrid-scale modelling based on the square of the velocity gradient tensor. *Flow, Turbulence and Combustion* 62: 183–200.

Pope, S.B. 2000. *Turbulent flows.* Cambridge, UK: Cambridge University Press.

Rider, W.J. 2007. Effective subgrid modeling from the ILES simulation of compressible turbulence. *Journal of Fluids Engineering* 129: 1493–1496.

Roux, A., Gicquel, L.Y.M., Sommerer, Y., and Poinsot, T.J. 2008. Large eddy simulation of mean and oscillating flow in a side-dump ramjet combustor. *Combustion and Flame* 152: 154–176.

Rubio, G., De Roeck, W., Baelmans, M., and Desmet, W. 2006. Numerical identi-fication of flow-induced oscillation modes in rectangular cavities using large eddy simulation. *International Journal for Numerical Methods in Fluids* 53: 851–866.

Sagaut, P. 2006. *Large eddy simulation for incompressible flows: An introduction.* Berlin, Germany: Springer.

Shu, C.W. and Osher, S. 1988. Efficient implementation of essentially non-oscillatory shock-capturing schemes. *Journal of Computational Physics* 77: 439–471.

Smagorinsky, J. 1963. General circulation experiments with the primitive equa-tions. *Monthly Weather Review* 91: 99–164.

Zóltak, J. and Drikakis, D. 1998. Hybrid upwind methods for the simulation of unsteady shock-wave diffraction over a cylinder. *Computer Methods in Applied Mechanics and Engineering* 162: 165–185.

CHAPTER 8

Further Topics and Challenges in DNS and LES

T
HE COMPLEX BEHAVIOR OF FLUID flow, including turbulence, can be mathematically considered as the consequence of a fairly simple set of equations as those presented in Chapter 1. Unfortunately, analytical solutions to even the simplest turbulent flows have not been found so far or they simply do not exist. However, a complete turbulent flow where the flow variables are functions of space and time can be obtained numerically. This is known as the CFD approach. As advanced CFD techniques, DNS and LES have been evolving rapidly over the last few decades, mainly as tools for research. DNS has emerged as the most powerful numerical tool to understand the fundamentals of flow instabilities, transition to turbulence, and relatively low or moderate Reynolds number turbulent flows, but high Reynolds number flows and large-scale problems remain untouchable. In addition, complex geometry problems still represent a significant difficulty for DNS. In the meantime, LES has been gradually evolving from a research tool toward a useful tool for practical applications.

DNS represents a methodology that avoids modeling or approximation of turbulence. The main purpose of DNS is to solve, to the best of our ability, for the turbulent flow field directly using highly accurate numerical techniques without modeling of the turbulence, on the platform of fast, large memory capacity computers. DNS means that the Navier–Stokes equations for fluid must be solved exactly, which is not a simple task. For the discretization, very fine grids are necessary. Any DNS code is very time consuming and has extensive storage requirements. Due to the continuously increasing computer power and availability of parallel computations, the contributions of DNS to turbulence research have been impressive and the future seems very promising. DNS offers advantages such as its great accuracy and the stringent control of the flow being investigated. However, the significantly high numerical fidelity required by DNS has to be maintained all the time, which provides a significant challenge for complex geometry problems. Another challenge of DNS is always related to its high computational costs. Although the Reynolds numbers of the simple turbulent flows investigated by DNS are currently approaching those of the small-scale experiments and DNS using hundreds of millions of grid points such as 512^3 can now be achieved on many supercomputers using parallel computations, there is still a long way to go for DNS to be directly applicable to most practical problems. Nevertheless, DNS has proven to be a very useful research tool that is able to provide results that are not possible using any other means. Due to its high accuracy, DNS can be used to perform "numerical experiments," to create simplified situations that are not possible in an experimental facility, and to isolate specific phenomena in the fluid flow. As the most accurate CFD methodology, DNS can be used to perform controlled studies that allow better insight into the fluid flow and allow scaling laws and turbulent models to be developed. The DNS databases offer the opportunity to extract information from turbulent flow fields, which cannot, or only with much difficulty, be obtained from experiments. The availability of a DNS database allows testing of the concepts behind models and may result in novel approaches to model turbulent flows. For instance, DNS can be used to evaluate the subgrid scale models in LES. The availability of detailed DNS flow information can also improve the understanding of the physical processes in turbulent flows, which can be used to develop various flow control strategies in practical applications.

The numerical techniques used in DNS are typically finite difference schemes, or a combination of spectral and finite difference schemes, although approaches such as spectral volume methods for complex geometries are also being explored. The main technical challenge of DNS remains the memory and computational speed requirements. DNS of the air flow past a complete airfoil would require a computer with 10^{18} flops capacity to be practical, which is far more than the currently available supercomputers with tera-flops (10^{12}). Most practical engineering problems such as the flow around a vehicle have too broad a range of scales to be directly computed using DNS. In addition, for chemically reacting flows encountered in combustion applications, the enormous computational requirements of DNS, including detailed chemical reaction schemes, are even more difficult to meet. Multiphase flow systems also represent a massive challenge for DNS. Apart from the difficulties associated with a higher Reynolds number, larger problem domains, and more complex flow physics, it is also difficult for DNS to deal with complex geometries where the accuracy of the numerics may be far from sufficient to represent the flow "exactly."

For most high Reynolds number flows encountered in practical applications, approximations such as LES, which computes only the large energy-containing scales, are more prevalent than DNS. Compared with the challenges of computer capacity and the numerics for DNS, the challenges of LES are not only the numerical issues, but also the physical issues associated with the subgrid-scale modeling since small scales need to be modeled. In LES, the effect of high-frequency velocity fluctuations within a flow field is estimated using modeling techniques, which inevitably leads to challenges when the flow is of a complex nature in physics.

Turbulent flows consist of a broad range of time and length scales. DNS attempts to solve all the relevant time and length scales, while LES tries to solve the large scales and model the small scales. A turbulent flow field is a typical multiscale flow system. As an interdisciplinary research field, multiscale modeling has emerged over the last few years in many areas of engineering and the physical sciences. Multiscale modeling is the field of solving physical problems that have important features at multiple scales, particularly multiple spatial scales. A broad range of scientific and engineering problems involve multiple scales. There has been a growing need to develop systematic modeling and simulation approaches for multiscale problems. The launch of the SIAM (Society for Industrial and Applied Mathematics) journal *Multiscale Modeling and Simulation* in 2003 indicated the formation of the research field. Since turbulent flows are typical multiscale problems, DNS and LES of turbulent flows can be discussed in the framework of the newly emerging field of multiscale modeling and simulation.

This chapter is devoted to further topics that are relevant to direct and large-eddy simulations and the challenges faced by DNS and LES. First, the important multiscale flow simulation is discussed, with an example given on simulations of turbulent atomization performed by Desjardins et al. (2008). Second, some challenges of DNS and LES are discussed, including

DNS of complex geometry problems and LES of complex physics problems such as multiphase and reacting flows. Finally, hybrid methods that are important to practical applications are discussed with examples of application, including detached eddy simulation (DES) and delayed DES (DDES).

I. MULTISCALE FLOW SIMULATIONS

A. The Concept of Multiscale Modeling

It is not an exaggeration to say that almost all problems have multiple scales. A broad range of scientific and engineering problems depend crucially on behavior at multiple scales, including the dynamics of biomolecules, the microstructure of materials, image and data analysis, earthquake physics, and atmospheric and oceanic dynamics. Well-known examples of fluid flow problems with multiple length scales include all types of turbulent flows, combustion in gas-turbine combustors and internal combustion engines, and vortical structures on the weather forecasting map. Even though multiscale problems have long been studied in mathematics, the current rapid development and the formation of a special research field of multiscale modeling and simulation are driven primarily by the use of mathematical models in engineering and physical sciences, particularly in material sciences such as polymers, chemistry, fluid dynamics, and biology. Problems in these areas are often multiphysics in nature; namely, the processes at different scales are governed by physical laws of different characters—for example, quantum mechanics at one scale and classical mechanics at another. Multiscale modeling and computation is a rapidly evolving research field that has a fundamental impact on computational science and applied mathematics and influences the relation between mathematics and engineering and physical sciences. Emerging from this research field is a need for new mathematics and new ways of interacting with mathematics for engineering and physical sciences. Fields such as mathematical physics and stochastic processes, which have so far remained in the background as far as modeling and computation is concerned, are moving to the frontier. There are several reasons for the rapid development of this research field. First, modeling at the level of a single scale, such as molecular dynamics or continuum theory, is becoming relatively mature. Second, the available computational capability has reached the stage when serious multiscale problems can be contemplated. Third, there is an urgent need from many other subjects of science and technology. For instance, nanoscience is a good example for the application of multiscale modeling techniques. In a multiscale problem, different physical laws may be required to describe the system at different scales. Take the example of fluids. At the macroscale such as meters or millimeters, fluids are accurately described by the density, velocity, and temperature fields, which obey the continuum Navier–Stokes equations. On the scale of the mean free path, it is necessary to use kinetic theory based on Boltzmann's equation. At the nanometer scale, molecular dynamics in the form of Newton's law has to be used to give the actual position and velocity of each individual atom that makes up the fluid. Multiscale modeling and simulation has its own special methods, such as the heterogeneous multiscale method, which was presented as a general methodology for an efficient numerical computation of problems with multiple scales by Weinan et al. (2003). For a fluid, moving from atomic level to macroscopic level, the theories for modeling and computation change from quantum mechanics described by the Schrödinger equation, to the molecular dynamics described by Newton's equation, and then to the kinetic theory described by Boltzmann's equation, followed by the continuum theory described by Navier–Stokes equations, which are the equations DNS and LES are based upon.

Traditional monoscale approaches have proven to be inadequate for many problems, even with the largest supercomputers, because of the range of scales and the prohibitively large number of variables involved. Thus, there is a growing need to develop systematic modeling and simulation approaches for multiscale problems. For multiscale problems, modeling and analysis across scales and multiscale algorithms are the key elements. In a multiscale problem, the boundaries between different levels of theories may vary, depending on the system being studied, but the overall trend is that a more detailed theory has to be used at each finer scale, giving rise to more detailed information on the system. There is a long history in mathematics for the study of multiscale problems. Fourier analysis has long been used as a way of representing functions according to their components at different scales. More recently, this multiscale multiresolution representation has been made much more efficient through wavelets. Another example of multiscale methods is the proper orthogonal decomposition technique discussed in Chapter 5. On the computational side, several important classes of numerical methods have been developed that address explicitly the multiscale nature of the solutions. As Weinan and Engquist (2003) summarized, these include multigrid methods, domain decomposition methods, fast multipole methods, adaptive mesh refinement techniques, and multiresolution methods using wavelets. All these methods can be used in CFD. From a modern perspective, the computational techniques described above are aimed at efficient representation or solution of the fine-scale problem. For many practical problems, full representation or solution of the fine-scale problem is simply impossible for the foreseeable future because of the overwhelming costs. Therefore, alternative approaches that are more efficient need to be adopted. A classical approach is to derive effective models at the scale of interest. Examples of such a technique are the RANS and LES modeling approaches. Certainly the concept of multiscale modeling and simulation is relevant to CFD, including RANS, LES, and DNS.

The modeling elements in RANS and LES are obvious. Although DNS is intended as model-free, it is not possible to achieve such a state in complex physics flows such as reacting flows and multiphase flows. In reacting flows, the chemistry of the combustion needs to be modeled so that it can be incorporated into the solver of the fluid flow at an affordable cost. For multiphase flows, the interaction between the different phases needs to be modeled, and mathematical models are also needed to track the interface between different phases. An example of such turbulent atomization is presented next to illustrate the application of multiscale modeling to fluid flow problems.

B. An Example of Multiscale Flow Modeling: Turbulent Atomization

The recent work by Desjardins et al. (2008) represents an example of multiscale flow modeling and simulation, where a gas-liquid two-phase flow system is investigated focusing on the liquid atomization in a turbulent flow environment. The breakup and atomization of liquid jets have a broad range of practical applications, including many industrial processes such as fuel injection in combustors, two-phase flow chemical reactors, spray coating, inkjet printing, and spray formations in medical applications. Sprays involving liquid- and gas-phase flows are widely utilized to provide rapid mixing between the liquid and its ambient environment. In most practical applications, the spray flow originated from an atomizer, which is often in the form of a jet, rapidly disintegrates into ligaments and further into droplets. This process of liquid jet breakup and atomization normally occurs near the nozzle orifice, and the flow develops into sprays at further downstream locations. The liquid disintegration is caused either by intrinsic (e.g., potential) or extrinsic (e.g., kinetic) energy, and the liquid is atomized either due to the kinetic energy contained in the liquid itself, by the interaction of the liquid sheet or jet with a (high-velocity) gas, or by means of mechanical energy delivered externally (e.g., by rotating devices). Although the liquid breakup and atomization process is an essential stage in the development of spray flows, it is not fully understood, particularly for high-speed jets. For the multiscale gas-liquid two-phase flow problem investigated by Desjardins et al. (2008), the modeling issues include the representation of the surface tension term as well as the density and viscosity jumps on the interface, while the challenging numerical issues include the accurate capturing of the gas-liquid interface.

Desjardins et al. (2008) presented a method for simulating incompressible two-phase flows by improving the conservative level set technique introduced by Olsson and Kreiss (2005). The method was then applied to simulate turbulent atomization of a liquid diesel jet at Re = 3000. The turbulent atomization problem investigated is physically very complex, involving momentum transfer between the two phases where the fine liquid droplets can be smaller than the grid size and the large scales include the liquid jet penetration and spreading. In the gas-liquid two-phase flow, surface instabilities, ligament formation, ligament stretching and fragmentation, and droplet coalescence all interact with turbulence to transform large-scale coherent liquid structures into small-scale droplets. There are several severe difficulties to numerically investigate such a complex physics problem.

The first difficulty is the large change in the material properties of the two phases; for example, the density and viscosity are significantly different in the two phases. In a diesel fuel injection, the liquid-to-gas density ratio can be as high as 40 while the viscosity ratio can be of the order of 30, which can move up to several hundred for aircraft engines. This large change in fluid properties corresponds to sharp gradients in the flow field, leading to severe numerical difficulties. In addition, the surface tension force on the gas-liquid interface needs to be mathematically and numerically represented, which also requires accurate localization and transport of the interface. Moreover, in the case of incompressible flows, the interface transport and localization should ensure that the volume of each phase is exactly conserved. As a multiscale problem, there is also a challenge coming from the small scales that the atomization process produces. In a numerical simulation, the solver normally generates liquid structures at the limit of numerical resolution. The formation of small liquid structures requires high numerical resolution to tackle.

For the modeling of gas-liquid two-phase flows, the volume of fluid (VOF) method has been broadly used, but the gas-liquid interface needs to be reconstructed from the VOF results and the exact location of the interface is unknown without this reconstruction (Scardovelli and Zaleski 1999). The front-tracking approach was introduced by Unverdi and Tryggvason (1992) and consists of discretizing the interface using an unstructured moving mesh that is transported in a Lagrangian fashion. However, the method requires frequent mesh rearrangements that affect the conservation of the liquid volume. The main limitation of this approach is the lack of automatic topology modification. Moreover, the parallelization of such a method for massively parallel computation is very challenging. Over the last several years, the level set method aiming at representing the interface implicitly by an isolevel of a smooth function, as described by Osher and Fedkiw (2003), has drawn significant attention in the field of interface modeling. Simple Eulerian scalar transport schemes can be used to transport this smooth function, and therefore highly accurate methods are available. Furthermore, the smoothness of the level set function makes the interface normals and curvature readily available for the surface tension calculation, while parallelization is straightforward and highly efficient. However, level set methods are typically plagued by mass conservation issues since no inherent conservation property of the level set function exists.

In an effort to reduce mass conservation errors while retaining the simplicity of the original method, Olsson and Kreiss (2005) and Olsson et al. (2007) proposed a simple modification to the level set method. By replacing the usual signed distance function of the classical level set approach with a hyperbolic tangent profile that is transported and reinitialized using conservative equations, Olsson and Kreiss (2005) showed that the mass conservation errors could be reduced by an order of magnitude in comparison with the results obtained with a signed distance function. Based on the work by Olsson and Kreiss (2005) and Olsson et al. (2007), Desjardins et al. (2008) made a few modifications to the level set method and presented the accurate conservative level set (ACLS) method, resulting in both improved accuracy and robustness.

Numerical simulations of liquid jet/sheet breakup and atomization in a gaseous atmosphere are very scarce so far, mainly due to the complex spatially developing nature of the flow and the fact that often high density ratios and capillary forces lead to serious numerical problems (Klein 2005). The surface tension force in the gas-liquid two-phase flow system needs to be modeled accurately. A commonly used approach is the continuum surface force (CSF) model developed by Brackbill et al. (1992). However, the

CSF model spreads out both the density jump and the surface tension force over a few cells surrounding the interface in order to facilitate the numerical discretization. Consequently, this approach tends to misrepresent the smallest front structures. For the handling of the large density ratio and the surface tension force in a multiphase flow solver, the ghost fluid method (GFM) as described by Fedkiw et al. (1999) provides a very attractive way in the context of finite differences, by using generalized Taylor series expansions that directly include these discontinuities. Since the GFM explicitly deals with the density jump, the resulting discretization is not affected. Similarly, the surface tension force can be included directly in the form of a pressure jump, providing an adequate sharp numerical treatment of this singular term. Accordingly, Desjardins et al. (2008) used the GFM for the surface tension term as well as for the density jump. However, the CSF model was still used for the discretization of the viscous terms due to the complexity involved in using the GFM for the viscous term. Their argument was that the viscous contribution is small in comparison with the convective terms in a turbulent flow, which is valid for high-speed flows involved in liquid atomization.

The level set approach is at the center of the gas-liquid interface modeling by Desjardins et al. (2008). In the level set approach, the interface is defined implicitly as an isosurface of a smooth function ϕ. This approach benefits from many advantages, including automatic handling of topology changes, efficient parallelization, as well as easy and accurate access to the interface normals and curvature. There are two different level set functions that may be used: the commonly used distance function proposed by Chopp (1993), and the hyperbolic tangent function that was used by Olsson and Kreiss (2005) in the context of their conservative level set method. In the level set method, the transport of the interface can simply be described by

$$\frac{\partial \phi}{\partial t} + \mathbf{u} \bullet \nabla \phi = 0 \tag{8.1}$$

In Equaton (8.1), $\mathbf{u}$ represents the velocity field. The classical level set technique by Chopp (1993) relies on representing the interface implicitly as the zero level set of a smooth function chosen to be the signed distance from the interface, that is,

$$|\phi(\mathbf{x},t)| = |\mathbf{x} - \mathbf{x}_\Gamma| \tag{8.2}$$

In Equation (8.2), x_Γ corresponds to the closest point on the interface from x, and $\phi(x,t) > 0$ on one side of the interface while $\phi(x,t) < 0$ on the other side. With this definition of the level set function, the interface itself corresponds to the $\phi(x,t) = 0$ isosurface. This choice leads to a very smooth ϕ-field, which can be adequately transported and differentiated to compute the normal vector n and the curvature κ of the interface defined as

$$n = \frac{\nabla\phi}{|\nabla\phi|} \quad \text{and} \quad \kappa = -\nabla \bullet n \tag{8.3}$$

In the classical level set technique, transporting the interface using Equation (8.1) leads to distortion in the level set function with the smoothness of ϕ lost, consequently leading to numerical problems. In order to ensure that ϕ remains smooth, an additional treatment is introduced to reshape ϕ into a distance function. This reinitialization of the distance profile can be performed using different procedures. The commonly used method is to solve a Hamilton-Jacobi equation as given by Sussman et al. (1994):

$$\frac{\partial\phi}{\partial\tau} + S(|\nabla\phi| - 1) = 0 \tag{8.4}$$

In Equation (8.4), S is a modified sign function as given in Peng et al. (1999), and τ represents a pseudotime. This equation can be discretized with high accuracy, therefore leading to an accurate reconstruction of the distance profile. However, the distance function level set approach in the context of multiphase flows can be problematic because neither the level set transport nor the reinitialization inherently conserves the volume of the region enclosed by the zero level set. For gas–liquid two-phase flows, this can lead to gains or losses in the mass of the liquid, which can lead to substantial errors in the numerical simulations.

In order to tackle the problem of liquid volume conservation, Olsson and Kreiss (2005) and Olsson et al. (2007) employed a hyperbolic tangent function ψ instead of the signed distance function ϕ. The hyperbolic tangent function ψ is defined as

$$\psi(x,t) = \frac{1}{2}\left\{\tanh\left[\frac{\phi(x,t)}{2\varepsilon}\right] + 1\right\} \tag{8.5}$$

In Equation (8.5), ε is a parameter that sets the thickness of the profile. Rather than defining the interface location by the isosurface $\phi = 0$, it is now defined by the location of the $\psi = 0.5$ isosurface. The transport of the interface can still be performed by solving the same equation as Equation (8.1) for ψ. However, it can also be written in conservative form provided the velocity field $\mathbf{u}$ is solenoidal, that is, $\nabla \bullet \mathbf{u} = 0$, namely,

$$\frac{\partial \psi}{\partial t} + \nabla \bullet (\mathbf{u}\psi) = 0 \tag{8.6}$$

With the level set transport equation written in conservative form, and the given definition of ψ, it is clear that the scalar ψ should be a conserved quantity. As in the case of the level set function ϕ, nothing ensures that solving Equation (8.6) preserves the form of the hyperbolic tangent profile ψ. As a result, an additional reinitialization equation needs to be introduced to reestablish the shape of the profile. As in Olsson et al. (2007), this equation can be written as

$$\frac{\partial \psi}{\partial \tau} + \nabla \bullet [\psi(1-\psi)\mathbf{n}] = \nabla \bullet [\varepsilon(\nabla \psi \bullet \mathbf{n})\mathbf{n}] \tag{8.7}$$

Equation (8.7) can be advanced in pseudotime τ and it consists of a compression term on the left-hand side that aims at sharpening the profile and of a diffusion term on the right-hand side that ensures that the profile remains of the characteristic thickness ε, and is therefore resolvable on a given mesh. It should be noted that this equation is also written in conservative form. As a result, solving successively for Equations (8.6) and (8.7) should accomplish the transport of the $\psi = 0.5$ isosurface, preserve the shape of the hyperbolic tangent profile, and ensure the conservation of ψ.

Based on the method of Olsson and Kreiss (2005) and Olsson et al. (2007), Desjardins et al. (2008) presented the accurate conservative level set (ACLS) method and the ACLS solution procedure can be briefly given as follows: (1) Advance the ψ field by solving Equation (8.6) using a semi-implicit Crank–Nicolson time integration. (2) Use a fast marching method (Desjardins et al. 2008) to efficiently reconstruct ϕ from ψ. (3) Compute the face normals from ϕ. (4) Compute the least-squares curvature from ϕ. (5) Perform the conservative reinitialization step: using a semi-implicit Crank–Nicolson time integration, Equation (8.7) is advanced.

In the simulations of the turbulent atomization performed by Desjardins et al. (2008), the ACLS method provides the details of the gas–liquid interface, where the material properties including density and viscosity are subject to a jump, while the velocity is continuous across the interface. The important change across the interface occurs to the variable "pressure" (Desjardins et al. 2008), which includes the surface tension term. In the solver, the ACLS procedure is coupled with the incompressible Navier–Stokes equations for the two-phase fluid flow. The full numerical solution procedure of Desjardins et al. (2008) can be summarized as follows:

- Using the ACLS methodology, advance the interface implicitly from $t^{n-1/2}$ to $t^{n+1/2}$ using the velocity at t^n.

- Advance the velocity field implicitly from t^n to t^{n+1} by solving the Navier–Stokes momentum equations without pressure gradient.

- Project the velocity field by solving the Poisson equation, making use of GFM. The solution of the pressure equation is computed using a Krylov-based method (van der Vorst 2003), preconditioned by a multigrid solver (Falgout and Yang 2002).

- Correct the velocity at t^{n+1} using the pressure gradient, again using GFM.

Sample results are shown in Figures 8.1 and 8.2. Figure 8.1 shows the instantaneous snapshots of the interface at different time instants. The interface displays a complex, turbulent behavior, as the liquid jet undergoes turbulent atomization. Many complex phenomena interact, leading to a

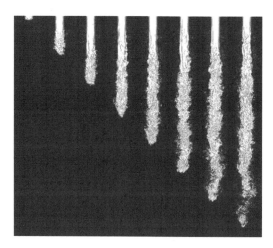

FIGURE 8.1 Turbulent atomization of a liquid diesel jet with $\Delta t = 2.5''$ between each image. (Desjardins et al. 2008; with permission from Elsevier Science Ltd.)

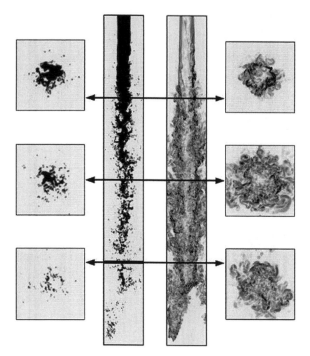

FIGURE 8.2 ψ-field (left) and magnitude of the vorticity (right) on two-dimensional axial and lateral cuts at $t = 22.8''$ for the turbulent liquid jet case. (Desjardins et al. 2008; with permission from Elsevier Science Ltd.)

fast break-up of the liquid core into ligaments and sheets, then droplets. It is interesting to note that by the end of the computational domain, the liquid core has fully disintegrated. The ψ-field as well as the magnitude of the vorticity field are presented in several two-dimensional cuts in Figure 8.2. The fully developed nature of the turbulence appears clearly, along with the chaotic nature of the interface. The flow appears to be vortical with complex fine scales. The results indicate that the numerical algorithm is robust for such a complex, turbulent, three-dimensional, multiphase, and multiscale flow problem.

II. CHALLENGES IN DNS AND LES: COMPLEX GEOMETRY AND SGS MODELING

A. Challenges in DNS: Complex Geometry

In general, there are many challenges in CFD, but problems associated with turbulence, shocks, and complex geometries stand out. As the most accurate CFD method, DNS has been developing rapidly over the last two decades.

For applications of DNS to larger domain and higher Reynolds number flows of practical relevance, the main technical challenges remain the memory and computational speed requirements. For complex physics flows such as chemically reacting flows and multiphase flows, the challenges include both the memory and computational speed requirements and efficient modeling or mathematical representation of the complex flow physics. Among various challenges for DNS, the most significant one in terms of numerical techniques is the treatment of complex geometries, where the accuracy of the numerical methods must be high enough, which is difficult to achieve. Apart from some special cases (e.g., Moulinec et al. 2004), DNS of complex geometry problems always represents a significant difficulty.

The field of DNS of turbulence has developed rapidly since the early simulations of homogeneous turbulence using spectral methods. Currently, high-order finite difference schemes and Chebyshev spectral methods are widely used in DNS in simple and separable computational domains. In the meantime, the computation of compressible flows in the presence of shocks has been based almost exclusively on low-order methods, especially in the shock region. Similarly, the computation of flows in complex geometries has been primarily based on low-order finite volume methods. DNS of turbulence in complex-geometry domains can be successful only if high-order accurate methods are employed. Efficient high-order discretization could be the most effective means of making real progress. Spectral element and the *hp* version of finite element methods were introduced in CFD in the early 1980s to address the issue of complex geometry domains and obtain high-order accuracy in such domains, which provides great flexibility in discretization and very effective means for adaptive refinement strategies. There are three versions of finite element methods. The classical *h*-version achieves the accuracy by refining the mesh, while the *p*-version keeps the mesh fixed and the accuracy is achieved by increasing the degree *p*. The *hp* version properly combines both approaches. The *hp* version is a useful development of finite element methods. However, the *hp* finite element method, and any high-order methods in general, are not as robust as low-order methods. They are very sensitive to boundary conditions and geometric or data singularities, which are not as stable as low-order methods and do not preserve monotonicity in the presence of discontinuous solutions such as shocks or contact discontinuities. For these methods to be computationally competitive with their low-order counterparts, they have to maintain fast convergence as their computational complexity is higher compared to low-order methods. Progress has been made over the last several years to address the computational complexity of spectral element and *hp* methods, where great advances have also been made on the theoretical side with rigorous theorems on stability (Karniadakis and Sherwin 2005). Such high-order numerical discretization methods have found use in varied situations where simple domain geometries have allowed easy application of boundary conditions. Karniadakis and Sherwin (2005) also demonstrated that spectral elements can have more widespread applicability through the use of grids of irregular "spectral elements."

As discussed in Chapter 5, the spectral finite volume methods (Wang 2002) are currently being explored as a promising numerical method for complex geometries, which are another type of high-order accurate, conservative, and computationally efficient scheme. In the spectral finite volume method, cell-averaged data from each triangular or tetrahedral finite volume is used to reconstruct a high-order approximation in the spectral volume, while Riemann solvers are used to compute the fluxes at the spectral volume boundaries. Since it does not require information from neighboring cells to perform reconstruction, it can be potentially very efficient and accurate. Recently, Wang (2007) reviewed several high-order spectral finite volume methods based on unstructured grids for the compressible Euler and Navier–Stokes equations, where the spatial and temporal discretizations were treated separately. Sample computational results were shown to illustrate the capability of selected methods. These high-order methods are expected to be more efficient than low-order methods for problems requiring high accuracy, including DNS and LES.

In the spectral finite volume or spectral-volume (SV) method, a single nonsingular stencil that can be applied to all the cells in an unstructured grid needs to be found. The spectral volumes (SVs) are unstructured grids of cells, triangles in 2D, and tetrahedra in 3D. Each SV is then partitioned into a number of "structured" subcells, referred to as control volumes (CVs), that support a polynomial expansion of a desired degree of precision. The unknowns are now the cell averages over the CVs. The CVs can be polygons or polyhedra. In 3D, they can have nonplanar faces, which must be subdivided into planar facets in order to perform the required integrations. All the SVs are partitioned in a geometrically similar manner. Therefore, a single, universal reconstruction for all SVs can be obtained. Due to the symmetry of the partition, only a few distinct coefficients appear in the expansion in terms of the CV unknowns. A CV face that lies on an SV boundary will have a discontinuity on its two sides.

A Riemann solver is then necessary to compute the flux on that face. If the flux is a linear function of the unknowns, the flux integration can be performed analytically without invoking quadratures (Liu 1994), and the result can be expressed as a weighted sum of all the CV unknowns in the two SVs. If the flux is a nonlinear function of the unknowns, a quadrature approximation of the appropriate degree of precision is required. The conservative variable on one side of a quadrature point can again be expressed as a weighted sum of the CV unknowns in the SV on that side. Since the quadrature points belong to just a few symmetry groups, the total number of distinct weights that need to be stored is relatively small. The reconstruction within each spectral volume is continuous. Therefore, a linear flux over a CV face that lies in the interior of an SV can be evaluated directly, and the weights for each type of facet can be stored. For a nonlinear flux, a similar procedure can be carried out for each quadrature point.

In the spectral finite volume method, the most general form of a conservation law can be written as

$$\frac{\partial u}{\partial t} + \nabla * F = 0 \tag{8.8}$$

In Equation (8.8), the conservative variable u can be a scalar or a vector, and the generalized flux F can be a scalar, vector, or tensor. The term $\nabla * F$ represents the divergence, curl, or gradient of F, depending on the physical definition of u. Integrating Equation (8.8) over each CV, we obtain

$$\frac{d}{dt} \int_{V_{j,i}} u \, dV + \sum_{k=1}^{K} \int_{S_{k,j,i}} dS * F = 0 \tag{8.9}$$

In Equation (8.9), K is the number of faces, $V_{j,i}$ is the volume of the jth CV in the ith SV, and $S_{k,j,i}$ is the area of planar facet k bounding $V_{j,i}$. The unknowns are the volume averages of u, defined as

$$\bar{u}_{j,i} = \frac{1}{V_{j,i}} \int_{V_{j,i}} u \, dV \tag{8.10}$$

In the spectral finite volume method, the partitioning of each SV into CVs depends on the choice of basis functions for the reconstruction and the order of accuracy is one order higher than the reconstruction degree of precision. Liu et al. (2006) partition the SV into N CVs, so that the reconstruction involves only the inversion of a square matrix. For a complete polynomial basis, a reconstruction of degree of precision n requires a partition into at least N CVs given by (Liu et al. 2006)

$$N = \begin{cases} (n+1)(n+2)/2, & \text{in 2D} \\ (n+1)(n+2)(n+3)/6, & \text{in 3D} \end{cases} \tag{8.11}$$

In order to reconstruct u within each SV, a set of complete polynomials of nth degree of precision $\phi_l(\mathbf{r})$ need to be introduced and $u_i(\mathbf{r})$ in the ith SV can be expanded as

$$u_i(\mathbf{r}) = \sum_{l=1}^{N} c_{l,i} \phi_l(\mathbf{r}) \tag{8.12}$$

Equation (8.10) can then be written as

$$\bar{u}_{j,i} = \sum_{l=1}^{N} R_{jl,i} c_{l,i} \tag{8.13}$$

In Equation (8.13),

$$R_{jl,i} = \frac{1}{V_{j,i}} \int_{V_{j,i}} \phi_l(\mathbf{r}) dV \tag{8.14}$$

Liu and Vinokur (1998) gave the exact integrations of polynomials over arbitrary polygons or polyhedra in terms of the coordinates of vertices. The elimination of the coefficients $c_{l,i}$ from Equations (8.12) and (8.13) can be shown succinctly using matrix algebra (Liu et al. 2006), leading to

$$u_i(\mathbf{r}) = \phi^T(\mathbf{r}) c_i \tag{8.15}$$

and

$$\bar{u}_i = R_i c_i \tag{8.16}$$

In Equations (8.15) and (8.16), c_i, $\bar{u}_i$, and $\phi(\mathbf{r})$ stand for the algebraic vectors with components $c_{l,i}$, $\bar{u}_{j,i}$, and $\phi_l(\mathbf{r})$, respectively, and the superscript T represents transposition, while R_i represents the matrix with elements $R_{jl,i}$. Eliminating c_i from Equations (8.15) and (8.16), the following cardinal form can be obtained:

$$u_i(\mathbf{r}) = L_i(\mathbf{r}) \bar{u}_i \tag{8.17}$$

where

$$L_i(\mathbf{r}) = \phi^T(\mathbf{r}) R_i^{-1} \tag{8.18}$$

In Equations (8.18), the algebraic row vector $L_i(\mathbf{r})$ has components $L_{j,i}(\mathbf{r})$, which are known as shape functions or cardinal basis functions. In expanded form, Equation (8.17) can be written as

$$u_i(\mathbf{r}) = \sum_{j=1}^{N} L_{j,i}(\mathbf{r}) \bar{u}_{j,i} \tag{8.19}$$

For a facet k on an interior CV face, if F is a linear function of u, the flux integral in Equation (8.9) can be evaluated by simply integrating u over that facet. Substituting the expression given in Equation (8.19), the result as a weighted sum of the CV unknowns can be given as

$$\int_{S_{k,i}} u \, dS = S_{k,i} \sum_{j=1}^{N} m_{k,j} \bar{u}_{j,i} \quad \text{with} \quad m_{k,j} = \frac{1}{S_{k,i}} \int_{S_{k,i}} L_{j,i}(\mathbf{r}) \, dS \tag{8.20}$$

In Equation (8.20), the surface integrals of the shape functions per unit area are universal, irrespective of the SVs. There are only a few of these coefficients for each partition, which can be calculated exactly and stored in advance. For nonlinear flux functions, the flux integral is evaluated by an nth degree of precision quadrature approximation of the form

$$\int_{S_{k,i}} dS * F = S_{k,i} \sum_{q} w_q \mathbf{n} * F[u_i(\mathbf{r}_q)] \tag{8.21}$$

where the w_q is the known quadrature weight. Using Equation (8.19), $u_i(\mathbf{r}_q)$ can be evaluated as a weighted sum of the CV unknowns

$$u_i(\mathbf{r}_q) = \sum_{j} l_{q,j} \bar{u}_{j,i} \tag{8.22}$$

In Equation (8.22), the weights $l_{q,j} = L_{j,i}(\mathbf{r}_q)$ are the functional values of the shape functions at the quadrature point, which are also universal, irrespective of the SVs. There are also only a few of these coefficients, which can be calculated exactly and stored in advance.

For a CV face on an SV boundary, since u may be discontinuous, the flux is replaced by a Riemann flux given by Liu et al. (2006) as

$$\mathbf{n} * F[u(\mathbf{r})] = F_{Riem}[u_L(\mathbf{r}), u_R(\mathbf{r}), \mathbf{n}] \tag{8.23}$$

For linear flux functions, the surface integral of Equation (8.21) can be expressed as a weighted sum of the $\bar{u}$ in both SVs sharing the face. For non-linear flux functions, $\mathbf{n} * F$ in Equation (8.21) is replaced by Equation (8.23) with u_L and u_R evaluated at quadrature points using Equation (8.22).

In using the spectral finite volume method, one simply specifies variables for the conservative variable u (for instance, the variable is density for the continuity equation while it is the mass flux for the momentum equation) and the generalized flux F in Equation (8.8), and then follows the above discretization procedure. Apparently this spectral finite volume method is significantly different from the traditional finite difference or finite volume methods. For a DNS application, parallel computation is always needed. For the spectral finite volume method, there are several aspects of the data structure that can lead to a very efficient parallelizable code. The global grid data consists of face numberings, vertex numberings and locations, and cell numberings. The topology is specified by listing for each face its vertex numbers, in an order indicating its orientation, and the two adjacent cell numbers. In order to make use of the universal nature of the partitioning, all global cells can be mapped into a single standard SV. Each global face can have three possible orientations in the standard SV for 2D, and 12 for 3D. For each SV partition, the local CV connectivities are predetermined. This information and the corresponding weights $m_{k,j}$ or $l_{q,j}$ can be read in as input to the code. It is then possible to have a single code valid for 2D or 3D, with any desired order of accuracy (Liu et al. 2006). Liu et al. (2006) also pointed out that there is an aspect inherent in the spectral finite volume method that permits optimum use of cache memory, resulting in great computational efficiency on modern computers. The spectral volume method might be able to improve the capability of DNS and LES significantly since practical engineering problems are predominantly of complex geometries. However, for the SV method to be more broadly used, the computational efficiency and accuracy of the method with different orders of accuracy need to be systematically assessed and compared with the compact finite difference schemes with the same orders of accuracy.

B. Challenges in LES: SGS Modeling

LES is a CFD technique that lies between DNS and RANS in terms of physical modeling and numerical accuracy. As examples of the applications of DNS and LES to practical engineering flows, Rodi (2006) presented DNS and LES results of three engineering flows carried out in his research group. The first example, simulated by using both DNS and LES, was the flow in a low-pressure turbine cascade with wakes passing periodically through the cascade channel. In this situation, the attention was focused on the laminar to turbulent transition of the boundary layers on the blade surfaces. In the second example, LES of the flow past the Ahmed body, which is a car model with slant back, was presented. In spite of the fairly simple geometry, the flow around the model has many features of the complex, fully 3D flow around real cars. The third example, for which LES was presented, is the flow past a surface-mounted circular cylinder of height-to-diameter ratio of 2.5. In this case, complex 3D flow develops with interaction of various vortices behind the cylinder. Apparently the second and third cases were not achievable for DNS. By means of these examples, Rodi (2006) showed that complex turbulent flows of engineering relevance can be predicted realistically by DNS and LES, albeit at large costs. The DNS and/or LES methods are particularly suited and superior to RANS methods for situations where unsteadiness such as vortex shedding and large-scale structures dominate the flow.

Over the last several years, DNS has evolved into an important tool for studying transition mechanisms or for a basic understanding of the flow, while LES has gradually evolved into a tool for practical engineering applications. Benhamadouche and Laurence (2003) evaluated LES against RANS, and investigated the cross-flow in a staggered tube bundle using LES and a transient Reynolds stress transport model (RSTM) in 2D and 3D, with two levels of grid refinement. The numerical method was based on a finite volume approach on unstructured grids using a collocated arrangement for all the unknown variables. It was shown that the LES results on the fine mesh are comparable to DNS, and experiments and reasonable agreement were still achieved with a coarse mesh. For the physical problem investigated by Benhamadouche and Laurence (2003), the RSTM also produced satisfactory results in 3D, but the 2D RSTM produced unphysical results. Hanjalić (2005) provided a view of some developments and a perspective on the future role of the RANS approach in the computation of turbulent flows and heat transfer in competition with LES. It was argued that RANS can further play an important role especially in industrial and environmental computations. Hanjalić (2005) emphasized the recent developments in RANS, as well as their potential in hybrid approaches in combination with the LES strategy. Limitations in LES at high Reynolds and Rayleigh number flows and heat transfer were revisited and some hybrid RANS/LES routes were discussed. The potential of very-large-eddy simulations (VLES) of flows dominated by (pseudo)deterministic eddy structures was also discussed and illustrated in an example of very high Rayleigh number thermal convection.

In general, the vast amount of LES results available in the literature highlighted the success achieved by LES so far. However, there are still many unsolved issues in LES, particularly those associated with the sub-grid scale (SGS) modeling of complex physical flows such as multiphase and reacting flows. Bellan (2000) reviewed the LES in the context of liquid spray computation, where issues related to modeling both the droplet interaction with the carrier flow and the interaction among droplets were discussed. Particular attention was devoted to LES aspects, which are different from those of single-phase flows. These include the correct portrayal of the droplet interaction with small turbulent scales, the modeling of SGS stresses, SGS heat and SGS species fluxes, and the accurate representation in the carrier flow equations of the source terms associated with the presence of the droplets. It was pointed out that there were a remarkably small number of studies addressing the combination of crucial phenomena needed for the accurate description of sprays (i.e., anisotropy, inhomogeneity, and three-dimensionality at the small scale), without appealing to strictly single-phase SGS models.

The database generated by DNS can be used to test SGS models a priori, while an posteriori study can be used to evaluate the impact of the SGS model on the flow field development. LES of complex physics flows such as multiphase flows is always challenging. Okong'o and Bellan (2004) presented numerical results of a three-dimensional temporal mixing layer with evaporating droplets with a priori analysis, while Leboissetier et al. (2005) conducted a posteriori analysis. In these studies, the gas-phase equations were written in a Eulerian frame for two perfect gas species, including the carrier gas and vapor emanating from the droplets, while the liquid-phase equations were written in a Lagrangian frame. The effect of droplet evaporation on the gas phase is considered through mass, momentum, and energy source terms. The LES models include those for the subgrid-scale fluxes and the filtered source terms, which both need to be assessed. In the field of LES of gas–liquid or gas–solid particle flows, there are still several important issues that remain at the stage of work in progress: the modeling of the interaction of the small-scale turbulence with the droplets or particles; the modeling of the SGS stresses, heat, and species fluxes; and the modeling of the source terms in the LES equations. These issues must all be resolved prior to attempting a meaningful LES calculation. In addition, the coupling between the large and small scales needs to be reflected in the modeling procedure for an accurate representation of the spray flow.

Another type of complex physics flow is reacting flows. Combustion LES appeared in the literature only a little more than a decade ago. Pitsch (2006) recently reviewed LES of turbulent combustion. It was argued and demonstrated that LES clearly offers advantages that move the state of the art toward accurate and predictive simulations of turbulent combustion. Although much research has been carried out in recent years, many fundamental questions still have to be addressed to realize the full predictive potential of combustion LES. Many studies have been performed in a priori testing and simulations of academic configurations as well as practical combustion devices exploring the potential of combustion LES. However, little fundamental research has been done that goes beyond the methods typically applied in the Reynolds-averaged context. The linear eddy mixing (LEM) model discussed in Chapter 6 provides an attempt at combustion LES. Within the context of LES, the LEM approach can be used to model the small-scale processes ranging from the grid resolution down to the Kolmogorov scale or the smallest scales related to chemical reaction in reduced dimension, while the large scales of the flow are calculated directly from the LES equations of the motion with an appropriate coupling procedure. However, there is clearly a need for further development in terms of SGS modeling of turbulent combustion.

III. HYBRIDIZATION: DETACHED EDDY SIMULATION (DES)

The basic concept of hybrid methods is to combine the advantages of two (or more) methods, yielding an optimal solution at least for a special class of flows, and to afford predictions of high Reynolds number flows with reasonable computational efforts. Hybrid methods such as the detached eddy simulation (DES) proposed by Spalart et al. (1997) and Spalart (2000) have been attracting more attention recently and can be a useful approach in practical simulations of wall-bounded flows. In DES, the attached flow regions near the walls are distinguished from the separated flow regions with detached eddies. An example of DES formulation is given by Spalart et al. (1997). The model was originally formulated by replacing the distance function d in the Spalart–Allmaras (S–A) one-equation model (Spalart and Allmaras 1992) with a modified distance function

$$\tilde{d} = \min[d, C_{DES}\Delta] \qquad (8.24)$$

where C_{DES} is a constant and Δ is the largest dimension of the grid cell in question. This modification of the S-A model, while very simple in nature, changes the interpretation of the model substantially. This modified distance function causes the model to behave as a RANS model in regions close to walls, and in a Smagorinsky-like manner away from the walls. This is usually justified with arguments that the scale dependence of the model is made local rather than global, which is the key feature of an LES, and that dimensional analysis backs up this claim. This DES approach may be used with any turbulence model that has an appropriately defined turbulence length scale (distance in the S-A model) and is a sufficiently localized model. Different turbulence models may be used in the DES approach, where the model should facilitate the switch between LES and RANS. In practice, many implementations of the DES approach allow for regions to be explicitly designated as RANS or LES regions. Also, many implementations use different numerical schemes in the RANS regions and the LES regions, where upwinded differences are frequently used in the RANS regions while central differences are often employed in the LES regions.

As a hybrid method, the DES method combines RANS and LES. It means that, near solid boundaries, the governing equations work in the RANS mode where all turbulent stresses are modeled using the traditional RANS turbulence models, while far away from solid boundaries, the method switches to the LES mode. Note that pressure and velocity fields are time- or ensemble-averaged in the near-wall region. Therefore, the unsteady vortical structures in the near-wall region are not resolved directly and DES is not able to give detailed information on the near-wall dynamic structures. The near-wall flow is predicted by RANS with statistical turbulence models, whereas the detached flow region, including the large-scale unsteady vortical structures, are computed by LES. Recently, Spalart (2009) systematically reviewed the DES approach and it was pointed out that the principal weakness of DES is its response to ambiguous grids, where DES on a given grid can be less accurate than RANS on the same grid or DES on a coarser grid in some situations. Partial remedies have been found, yet dealing with thickening boundary layers and shallow separation bubbles still represents a great challenge for DES. Apart from the nonmonotonic response to grid refinement, DES also needs to deal with issues such as different numerical needs in the RANS and LES regions and the absence of a theoretical order of accuracy, and the coupling between the RANS and LES modeling. In order to overcome the problem of the nonmonotonic response to grid refinement of the DES, Spalart et al. (2006) proposed the delayed detached eddy simulation (DDES), which is one of the latest evolutions of the original DES approach. The originality of the DDES is the use of a function switching continuously from 0 in the boundary layer to 1 in the detached regions. For the DES approach to be an effective hybrid method, the numerical issues and the coupling between RANS and LES modeling still need to be further investigated.

Detached eddy simulation has found applications in many practical engineering problems. One such example is noise prediction. Noise is simply the pressure fluctuation in the flow field. The pressure fluctuations that can be sensed by the healthy ear are normally from slightly below 20 μPa (the nominal hearing threshold). The sound pressure level is a dimensional quantity and the measurement units are decibels (dB):

$$L_p^* = 10 \cdot \log_{10} \left(\frac{p_{rms}^{'*}}{20 \times 10^{-6}} \right)^2 \tag{8.25}$$

From nondimensional quantities, the dimensional sound pressure level (dB) can be calculated from the following equation:

$$L_p^* = 20 \cdot \log_{10} \frac{p_{rms}^{'} p_{ref}^*}{20 \times 10^{-6}} \tag{8.26}$$

Numerical prediction of the noise field is difficult for two reasons: (1) the computational domain has to be big enough and simulation has to be time-dependent for the acoustic predictions to be meaningful, and (2) the numerical methods need to be accurate enough so that the small acoustic energy can be accurately captured. The latter is particularly important because the acoustic energy is normally several orders of magnitude smaller than the flow energy (considering that the pressure fluctuation can be a few Pa while the pressure of flow is at least 10^5 Pa). Computational aeroacoustics is a typical multiscale simulation and modeling problem, where the small scales are related to the turbulence structures and the large scales can be related to the acoustic waves. Both DNS and LES can be used to predict the acoustic field, apart from the modeling approach—acoustic analogy methods that are applied mostly to reduce acoustic sound sources to simple emitter types. However, all these methods have their limitations. Due to the large domain required, DNS is often restricted to two-dimensional or axisymmetric simulations for aeroacoustic applications (e.g., Jiang et al. 2004; 2006). LES often becomes difficult due to the modeling of the near-wall flow region. Under this circumstance, DES becomes a viable option by combining RANS in the near-wall regions and LES in the main flow regions.

Terracol et al. (2006) performed a hybrid LES/RANS study of airframe noise prediction, which was focused on the first step of such hybrid methods. The unsteady aerodynamic noise sources were predicted by means of a 3D unsteady simulation of the flow. A zonal LES method based on the nonlinear disturbance equations (NLDE) was used for the numerical prediction of the aerodynamic noise sources. This method makes it possible to perform only zonal LES close to the main elements responsible for sound generation, while the overall configuration is treated only by a RANS approach. The zonal RANS/LES method was used to solve the set of the NLDE in order to reconstruct turbulent fluctuations around a given mean flow. The principle of the method is to decompose the conservative variables vector as a mean and a fluctuating part. Further details of the method can be found in Terracol et al. (2006). The flow over a thin flat plate ended by a blunted trailing edge is considered with four computational cases performed: (1) full LES—"FULL," (2) nonreflecting inflow—"LAM," (3) recycling perturbation treatment—"REC," and (4) analytical turbulent boundary layer (TBL) model. Figure 8.3 shows a three-dimensional view

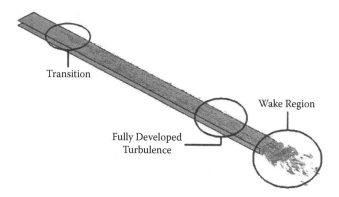

FIGURE 8.3 Three-dimensional view of the flow in the LES on the full configuration. (Terracol 2006; with permission from Springer-Verlag.)

FIGURE 8.4 Acoustic wave emission at the trailing edge. (Terracol 2006; with permission from Springer-Verlag.)

of the flow in the full configuration, therefore exhibiting its main physical features, including transition, fully developed boundary layers, and a 3D turbulent wake. As shown in Figure 8.4, the wake occurring at the trailing edge is responsible for the emission of an acoustic wave. Figures 8.5 and 8.6 compare the physical features of the flow obtained in each case to the reference LES flow. It is to be noted that in the case of "LAM," no turbulent structures are visible in the boundary layer, as might be expected. As a consequence, the level of three-dimensionality and turbulence in the wake looks slightly decreased in this case. However, for the two zonal simulations "REC" and "ANA," the flow looks very similar to the one obtained by the full LES and exhibits a highly three-dimensional behavior, with a good representation of the typical structures observed in TBL.

In another application, Terracol et al. (2005) presented hybrid methods for airframe noise numerical prediction. The three-dimensional,

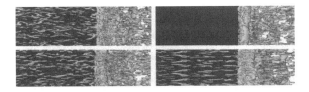

FIGURE 8.5 Top view of the flow. Top left: LES on the full configuration; top right: LES with nonreflecting inflow; bottom left: LES with an additional recycling treatment for the perturbation; bottom right: analytical turbulent boundary layer model. (Terracol 2006; with permission from Springer-Verlag.)

FIGURE 8.6   Side view of the flow. Top left: LES on the full configuration; top right: LES with nonreflecting inflow; bottom left: LES with an additional recycling treatment for the perturbation; bottom right: analytical turbulent boundary layer model. (Terracol 2006; with permission from Springer-Verlag.)

compressible, unsteady filtered Navier–Stokes equations are used to solve the flow of a Newtonian viscous fluid around a NACA0012 airfoil. The filtering operator, classically defined as a convolution product on the computational domain, is assumed to commute with time and spatial derivatives. The LES is based on the discretization of the compressible Navier–Stokes equations on multiblock structured meshes by a finite volume technique. The Navier–Stokes equations are discretized using a cell-centered finite volume technique and structured multiblock meshes. The viscous fluxes are discretized by a second-order accurate centered scheme. For efficiency reasons, implicit time integration is employed to deal with the very small grid size encountered near the wall. An approximate Newton method is employed to solve the nonlinear problem. At each iteration of this inner process, the inversion of the linear system relies on the lower-upper symmetric Gauss–Seidel implicit method.

The 3D curvilinear computational grid is obtained by replication in the spanwise direction y of a 2D curvilinear structured grid made of two domains. Domain #1 is located upstream of the C-shaped trailing edge (TE), with 309 points along the airfoil body and 97 points in the radial direction. Domain #2 is located downstream from the TE, with 227 points in the z direction (including 35 points on the TE bluntness) and 103 points in the x direction. The computational grid schematic is shown in Figure 8.7. A nonslip condition is applied at the airfoil surface and a periodic condition is imposed in the spanwise direction. Nonreflecting characteristic boundary conditions are applied for the far field. Moreover, a steady RANS computation using Baldwin-Lomax models provides an initial flow solution. It

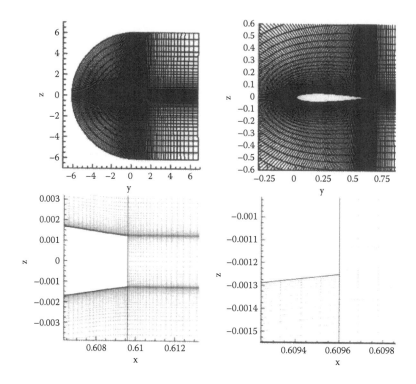

FIGURE 8.7 Schematic of the computational grid used for the airframe noise prediction. (Terracol et al. 2005; with permission from Springer-Verlag.)

should be noted that the LES becomes naturally three-dimensional in the regions of laminar-to-turbulent transition, without requiring any artificial numerical triggering.

Figure 8.8 shows details of the instantaneous (left) and time-averaged (right) flow streamlines around the blunted trailing edge. Figure 8.9 shows instantaneous isovalues of the pressure fluctuations inside the flow. At every point of the LES grid, pressure fluctuations are computed by sub-tracting the time-averaged pressure from the instantaneous pressure. Concentric waves are clearly observed near the TE, with a wavelength corresponding to the vortex shedding frequency. It is interesting to notice that the wave pattern corresponding to the vortex shedding noise vanishes at a half-chord from the airfoil, when, at the same time, larger wavelengths are observed much further on. This is explained by the radial stretching of the LES grid, which acts on the noise field as a low-pass filter. It is known

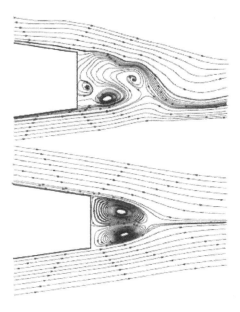

FIGURE 8.8 The instantaneous (above) and time-averaged (below) flow streamlines at the trailing edge. (Terracol et al. 2005; with permission from Springer-Verlag.)

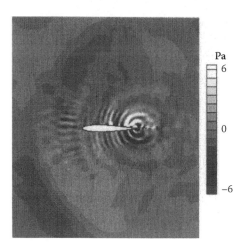

FIGURE 8.9 Instantaneous isovalues of pressure fluctuations obtained from LES data. (Terracol et al. 2005; with permission from Springer-Verlag.)

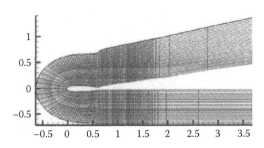

FIGURE 8.10   Final problem-adapted acoustic grid for the airframe noise prediction. (Terracol et al. 2005; with permission from Springer-Verlag.)

that the propagation of an acoustic wave will not be correctly simulated if it is discretized by using less than four or six cells per wavelength.

An acoustic grid was derived from the LES grid, following specific constraints: (1) the homogeneity of grid refinement and (2) the average cell size (with respect to the smallest wave lengths). This acoustic grid is shown in Figure 8.10. The interior border, on which LES data will be injected in the Euler domain, follows the airfoil surface at an average distance of 1% of the chord length. The outer border of the acoustic domain is approximately one chord away from the airfoil, since it was checked that the mean flow is quasi-uniform beyond this distance. Figure 8.11 presents a closer view of this grid, near the airfoil.

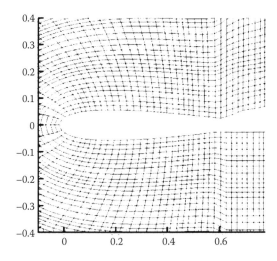

FIGURE 8.11   Close-up of the final problem-adapted acoustic grid. (Terracol et al. 2005; with permission from Springer-Verlag.)

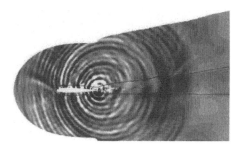

FIGURE 8.12 Coupled LES/Euler instantaneous pressure fluctuations. (Terracol et al. 2005; with permission from Springer-Verlag.)

Figure 8.12 shows isovalue contours of instantaneous pressure fluctuation field computed from (1) LES, inside the injection interface, and (2) propagation via Euler equations under a small perturbation hypothesis (from LES data injection) outside the injection interface. Figure 8.13 shows a closer view of Figure 8.12 centered on the airfoil. This view shows that there is no discontinuity at the injection interface between the LES wave fronts and the wave fronts predicted by the Euler equations under a small perturbation hypothesis.

Different from the DES approach combining RANS and LES, there is also a modeling approach that lies between the RANS and LES approaches. Speziale (1998) proposed such a very-large-eddy simulation (VLES) approach. In general, the simulation can be regarded as a VLES if the filter and grid are too coarse to resolve 80% of the energy.

FIGURE 8.13 Close-up of the coupled LES/Euler instantaneous pressure fluctuations. (Terracol et al. 2005; with permission from Springer-Verlag.)

This corresponds to a coarse-grid LES, which some have viewed as being equivalent to an unsteady RANS. VLES attempts to bridge the gap between the traditional RANS simulation and the traditional LES. This approach affords an intermediate resolution of turbulence scales relative to those of RANS and LES. In VLES, the very large scales of turbulence are directly calculated, and the effects of the unresolved scales are accounted for by an eddy viscosity model that is evolved from state-of-the-art models used in the RANS approach. The VLES is not a hybrid method like DES.

For the hybrid DES, although it seems to be a useful technique for the predictions of high Reynolds number, wall-bounded flows, a variety of open issues need to be addressed before one can rely on such a hybrid method. These include, in particular, the demand for appropriate coupling techniques between LES and RANS, adaptive control mechanisms, and proper SGS-RANS turbulence models. For the wall boundary treatment, the wall functions discussed in Chapter 2 are typical RANS CFD approaches in the near-wall region, which can be used in the hybrid method. In a hybrid DES approach, the quality of the numerical results depends on both the LES and RANS and their coupling. The final numerical results of DES rely on LES and RANS turbulence modeling, the numerics, the coupling between the two approaches, and the wall boundary conditions implemented in RANS modeling in the near-wall region.

REFERENCES

Bellan, J. 2000. Perspectives on large eddy simulations for sprays: issues and solutions. *Atomisation and Sprays* 10: 409–425.

Benhamadouche, S. and Laurence, D. 2003. LES, coarse LES, and transient RANS comparisons on the flow across a tube bundle. *International Journal of Heat and Fluid Flow* 24: 470–479.

Brackbill, J.U., Kothe, D.B., and Zemach, C. 1992. A continuum method for modeling surface tension. *Journal of Computational Physics* 100: 335–354.

Chopp, D.L. 1993. Computing minimal surfaces via level set curvature flow. *Journal of Computational Physics* 106: 77–91.

Desjardins, O., Moureau, V., and Pitsch, H. 2008. An accurate conservative level set/ghost fluid method for simulating turbulent atomization. *Journal of Computational Physics* 227: 8395–8416.

Falgout, R.D. and Yang, U.M. 2002. HYPRE: A library of high performance preconditioners. In *Computational Science—ICCS 2002 Part III*, eds. P.M.A. Sloot, C.J.K. Tan, J.J. Dongara, and A.G. Hoekstra, 632–641. Berlin, Germany: Springer-Verlag.

Fedkiw, R., Aslam, T., Merriman, B., and Osher, S. 1999. A non-oscillatory Eulerian approach to interfaces in multimaterial flows (the ghost fluid method). *Journal of Computational Physics* 152: 457–492.

Hanjalić, K. 2005. Will RANS Survive LES? A View of Perspectives. *ASME Journal of Fluids Engineering* 127: 831–839.

Jiang, X., Avital, E.J., and Luo, K. H. 2004. Sound generation by vortex pairing in subsonic axisymmetric jets. *AIAA Journal* 42: 241–248.

Jiang, X., Zhao, H., and Cao, L. 2006. Numerical simulations of the flow and sound fields of a heated axisymmetric pulsating jet. *Computers & Mathematics with Applications* 51: 643–660.

Karniadakis, G. and Sherwin, S. 2005. *Spectral/hp element methods for computational fluid dynamics*, 2nd ed. Oxford: Oxford University Press.

Klein, M. 2005. Direct numerical simulation of a spatially developing water sheet at moderate Reynolds number. *International Journal of Heat and Fluid Flow* 26: 722–731.

Leboissetier, A., Okong'o, N.A., and Bellan, J. 2005. Consistent large-eddy simulation of a temporal mixing layer laden with evaporating droplets. Part 2. A posteriori modeling. *Journal of Fluid Mechanics* 523: 37–78.

Liu, Y. 1994. A generalized finite volume algorithm for solving Maxwell's equations on arbitrary grids. In *Proceedings of 10th Annual Review of Progress in Applied Computational Electromagnetics*.

Liu, Y. and Vinokur, M. 1998. Exact integration of polynomials and symmetric quadrature formulas over arbitrary polyhedral grids, *Journal of Computational Physics* 140: 122–147.

Liu, Y., Vinokur, M., and Wang, Z.J. 2006. Spectral (finite) volume method for conservation laws on unstructured grids V: Extension to three-dimensional systems. *Journal of Computational Physics* 212: 454–472.

Moulinec, C., Hunt, J.C.R., and Nieuwstadt, F.T.M. 2004. Disappearing wakes and dispersion in numerically simulated flows through tube bundles. *Flow, Turbulence, and Combustion* 73: 95–116.

Okong'o N.A. and Bellan, J. 2004. Consistent large-eddy simulation of a temporal mixing layer laden with evaporating droplets. Part 1. Direct numerical simulation, formulation and a priori analysis. *Journal of Fluid Mechanics* 499: 1–47.

Olsson, E. and Kreiss, G. 2005. A conservative level set method for two phase flow. *Journal of Computational Physics* 210: 225–246.

Olsson, E., Kreiss, G., and Zahedi, S. 2007. A conservative level set method for two phase flow II. *Journal of Computational Physics* 225: 785–807.

Osher, S. and Fedkiw, R. 2003. *Level set methods and dynamic implicit interfaces*. New York: Springer.

Peng, D., Merriman, B., Osher, S., Zhao, H., and Kang, M. 1999. A PDE-based fast local level set method. *Journal of Computational Physics* 155: 410–438.

Pitsch, H. 2006. Large-eddy simulation of turbulent combustion. *Annual Review of Fluid Mechanics* 38: 453–482.

Rodi, W. 2006. DNS and LES of some engineering flows. *Fluid Dynamic Research* 38: 145–173.

Scardovelli, R. and Zaleski, S. 1999. Direct numerical simulation of free-surface and interfacial flow. *Annual Review of Fluid Mechanics* 31: 567–603.

Spalart, P.R. 2000. Strategies for turbulence modeling and simulations. *International Journal of Heat and Fluid Flow* 21: 252–263.

Spalart, P.R. 2009. Detached-eddy simulation. *Annual Review of Fluid Mechanics* 41: 181–202.

Spalart, P.R. and Allmaras, S.R. 1992. A one-equation turbulence model for aerodynamic flows. AIAA Paper 92-0439.

Spalart, P.R., Deck, S., Shur, M.L., Squires, K.D., Strelets, M.Kh., and Travin, A. 2006. A new version of detached-eddy simulation, resistant to ambiguous grid densities. *Theoretical Computational Fluid Dynamics* 20: 181–195.

Spalart, P.R., Jou, W.H., Strelets, M., and Allmaras, S.R. 1997. Comments on the feasibility of LES for wings and on a hybrid RANS/LES approach. In *Advances in DNS/LES*, ed. C. Liu and Z. Liu, 137–148. Columbus, OH: Greyden Press.

Speziale, C.G. 1998. Turbulence modeling for time-dependent RANS and VLES: A review. *AIAA Journal* 36: 173–183.

Sussman, M., Smereka, P., and Osher, S. 1994. A level set method for computing solutions to incompressible two-phase flow. *Journal of Computational Physics* 114: 146–159.

Terracol, M. 2006. A zonal RANS/LES approach for noise sources prediction. *Flow, Turbulence and Combustion* 77: 161–184.

Terracol, M., Manoha, E., Herrero, C., Labourasse, E., Redonnet, S., and Sagaut, P. 2005. Hybrid methods for airframe noise numerical prediction. *Theoretical and Computational Fluid Dynamics* 19: 197–227.

Unverdi, S. and Tryggvason, G. 1992. A front-tracking method for viscous incompressible multi-fluid flows. *Journal of Computational Physics* 100: 25–37.

van der Vorst, Henk A. 2003. *Iterative Krylov methods for large linear systems,* 1st ed. Cambridge, UK: Cambridge University Press.

Wang, Z.J. 2002. Spectral (finite) volume method for conservation laws on unstructured grids—basic formulation. *Journal of Computational Physics* 178: 210–251.

Wang, Z.J. 2007. High-order methods for the Euler and Navier-Stokes equations on unstructured grids. *Progress in Aerospace Sciences* 43: 1–41.

Weinan, E. and Engquist, B. 2003. Multiscale modeling and simulation. *Notices of the American Mathematical Society* 50: 1062–1070.

Weinan, E., Engquist, B., and Huang, Z.Y. 2003. Heterogeneous multiscale method: A general methodology for multiscale modeling. *Physics Review B* 67, Article Number: 092101.

Appendix: FORTRAN 90 Routines of the Finite Difference Schemes

Finite difference schemes are widely used in DNS and LES, which offers flexibility in the specification of boundary conditions. In addition, the parallelization of computer programs using finite difference schemes is relatively straightforward. The computing costs of finite difference schemes are also generally lower than finite volume methods. Since upwind-biased schemes inherently introduce some form of artificial smoothing or dissipation error that makes them inappropriate for long-time integration, central difference schemes that do not introduce artificial dissipation have been predominantly used in DNS and LES. Lele (1992) developed high-order accurate, narrow-stencil, finite difference schemes appropriate for problems with a wide range of scales, known as the "compact" or Padé schemes. These centered schemes developed by Lele (1992) require small stencil support, which are of particular interest in DNS. The main advantage of compact schemes is simplicity in boundary condition treatment and smaller truncation error compared to their noncompact counterparts of equivalent order, as described in Chapter 5. In this appendix, as an example of high-order finite difference schemes, Fortran 90 subroutines of the Padé 3/4/6 scheme are presented first. In this Padé 3/4/6 scheme, the formal accuracy of sixth order holds in the interior of the computational domain. The scheme is of third-order accuracy at the boundary points, of fourth-order at the next-to-the-boundary points, and of sixth-order at inner points only. Modular Fortran 90 subroutines on the Padé 3/4/6 scheme are given, including the subroutines that preserves the sixth-order accuracy at a symmetry boundary. Second, Fortran 90 subroutines on the second-order central differentiation are presented due to their wide applications in CFD including in LES, with two subroutines using a three-point stencil and a five-point stencil formulations respectively.

I. PADÉ 3/4/6 SCHEME

The Padé scheme has been widely used and now is the state-of-the-art numerical method in DNS codes for fluid flow and combustion problems. The main advantages of the Padé scheme are the low computing costs associated with the small-stencil support and the simplicity in boundary condition treatment. Lele (1992) systematically presented the Padé scheme. For the Padé scheme, at the boundary points or points near the boundaries, central differencing is not possible because points outside the computational domain cannot be included. Boundary closures of the Padé scheme inevitably lead to the use of lower-order schemes for points near the boundary. For the Padé 3/4/6 scheme discussed in Chapter 5 (also in Chapter 2), the formal accuracy of sixth order holds only in the interior of the computational domain. In this formulation, the number of neighboring points involved on one side of the boundary points or points near the boundaries is the same as that of the inner points. The scheme is of third-order accuracy at the boundary points, of fourth-order at the next-to-the-boundary points, and of sixth-order at inner points only. The formula for the Padé 3/4/6 scheme were given in Equations (2.17)–(2.20). The application of the Padé scheme also requires solving a tridiagonal matrix system. For the Padé 3/4/6 scheme, the tridiagonal matrix inversion can be conveniently achieved by the Gaussian elimination known as the tridiagonal matrix algorithm (TDMA), or Thomas algorithm (Conte and de Boor, 1972). In a practical simulation, nonuniform grids may be used to solve the flow field more efficiently. Grid transformation can be used to link the physical domain and the computational domain. For both the first- and second-order derivatives, once the derivatives in the computational domain on an equally spaced mesh are obtained, derivatives in the physical domain with possibly nonuniform grid distribution can be obtained using the metrics for grid transformation (Anderson 1995). As discussed in Chapter 2, the sixth-order accuracy of the Padé 3/4/6 scheme can be preserved at the symmetry boundary, by applying the symmetry conditions to both the primitive variables and their first- and second-order derivatives in the symmetry direction.

In the following, the Fortran 90 subroutines of the Padé 3/4/6 scheme in the computational domain are given. The subroutines consider a two-dimensional field and give the first and second derivatives of a discretized function f on a grid with the number of points represented by (nx, ny). Apparently adaptation of the subroutines to one-dimensional or three-dimensional problems is straightforward. In the following, subroutines, dx(f), dy (f), d2x(f), and d2y(f) give the output of the first and second derivatives in the x and y directions using the Padé 3/4/6 scheme with third-order boundary closures, respectively. In the meantime, subroutines dys(f) and d2ys(f) give the output of the first and second derivatives in the symmetric y direction preserving the sixth-order accuracy at the boundary $y = 0$ for variables without a sign change across the symmetry boundary (as given in Equation [2.15]) such as density ρ, while subroutines dyv(f) and d2yv(f) give the output of the first and second derivatives in the symmetric y direction preserving the sixth-order accuracy at the boundary $y = 0$ for variables with a sign change across the symmetry boundary (as given in Equation [2.16]) such as density y velocity component v. The modular subroutines given can be conveniently implemented into a Fortran code.

```
subroutine dx(f)
include 'incl.for'
dimension f(nx,ny)
call dpadex(f)
return
end
!
subroutine dy(f)
include 'incl.for'
dimension f(nx,ny)
call dpadey(f)
return
end
!
subroutine dys(f)
```

```fortran
include 'incl.for'
dimension f(nx,ny)
call dpadys(f)
return
end
!
subroutine dyv(f)
include 'incl.for'
dimension f(nx,ny)
call dpadyv(f)
return
end
!
subroutine d2x(f)
include 'incl.for'
dimension f(nx,ny)
call d2padx(f)
return
end
!
subroutine d2y(f)
include 'incl.for'
dimension f(nx,ny)
call d2pady(f)
return
end
!
subroutine d2ys(f)
include 'incl.for'
dimension f(nx,ny)
call d2pdys(f)
return
end
!
subroutine d2yv(f)
include 'incl.for'
dimension f(nx,ny)
call d2pdyv(f)
return
end
!
!------------------------------------------------------------
```

```fortran
! Subroutines for differentiation with Pade scheme in
! both x and y directions.
!-----------------------------------------------------------
subroutine dpadex(f)
!
!      Lele's Pade scheme 3/4/6 (6th order in interior)
!
include 'incl.for'
!
dimension f(nx,ny)
dimension a(nx),b(nx,ny),c(nx),d(nx,ny),cc(nx,ny)
dimension x(nx),hx(nx),h2x(nx)
common /pade/ a0,b0,c0,a1,b1,c1
common /padx/ ds,a,c
common /mapx/ x,hx,h2x
!
do j=1,ny
!
b(1,j)    = 2.
b(nx,j) = 2.
  d(1,j)  = (-5.*f(1,j) + 4.*f(2,j) + f(3,j))/ds
d(nx,j) = (5.*f(nx,j) - 4.*f(nx-1,j) - f(nx-2,j))/ds
!
b(2,j)      = 4.
b(nx-1,j) = 4.
d(2,j)      = 3.*(f(3,j) - f(1,j))/ds
d(nx-1,j) = 3.*(f(nx,j) - f(nx-2,j))/ds
!
do i=3,nx-2
b(i,j) = a0
d(i,j) = 0.5*b0*(f(i+1,j)-f(i-1,j))/ds
d(i,j) = d(i,j) + 0.25*c0*(f(i+2,j)-f(i-2,j))/ds
end do
end do
!
call thomas(nx,ny,a,c,b,d,cc)
!
do j=1,ny
do i=1,nx
f(i,j)=d(i,j)/hx(i)
end do
end do
```

```fortran
!
return
end
!-----------------------------------------------------------
subroutine dpadey(f)
!
!      Lele's Pade scheme 3/4/6 (6th order in interior)
!
include 'incl.for'
!
dimension f(nx,ny)
dimension a(ny),b(ny,nx),c(ny),d(ny,nx),cc(ny,nx)
dimension y(ny),hy(ny),h2y(ny)
common /pade/ a0,b0,c0,a1,b1,c1
common /pady/ ds,a,c
common /mapy/ y,hy,h2y
!
do i=1,nx
!
b(1,i)    = 2.
b(ny,i) = 2.
d(1,i)    = (-5.*f(i,1) + 4.*f(i,2) + f(i,3))/ds
d(ny,i) = (5.*f(i,ny) - 4.*f(i,ny-1) - f(i,ny-2))/ds
!
b(2,i)      = 4.
b(ny-1,i) = 4.
d(2,i)      = 3.*(f(i,3) - f(i,1))/ds
d(ny-1,i) = 3.*(f(i,ny) - f(i,ny-2))/ds
!
do j=3,ny-2
b(j,i) = a0
d(j,i) = 0.5*b0*(f(i,j+1)-f(i,j-1))/ds
d(j,i) = d(j,i) + 0.25*c0*(f(i,j+2)-f(i,j-2))/ds
end do
end do
!
call thomas(ny,nx,a,c,b,d,cc)
!
do j=1,ny
do i=1,nx
  f(i,j)=d(j,i)/hy(j)
end do
```

```fortran
end do
!
return
end
!-----------------------------------------------------------
subroutine dpadys(f)
!
!       Lele's Pade scheme 3/4/6 (6th order in interior)
!       for "twin" variables [phi(+y)=phi(-y)] of
symmetrical domain
!       (6th order at y=0)
!
include 'incl.for'
!
dimension f(nx,ny)
dimension a(ny),b(ny,nx),c(ny),d(ny,nx),cc(ny,nx),cys(ny)
dimension y(ny),hy(ny),h2y(ny)
common /pade/ a0,b0,c0,a1,b1,c1
common /pady/ ds,a,c
common /mapy/ y,hy,h2y
!
do i=1,nx
!
b(1,i) = a0
b(ny,i) = 2.
d(1,i)    = 0.5*b0*(f(i,2)-f(i,2))/ds
d(1,i)    = d(1,i) + 0.25*c0*(f(i,3)-f(i,3))/ds
d(ny,i) = (5.*f(i,ny) - 4.*f(i,ny-1) - f(i,ny-2))/ds
!
b(2,i)       = a0
b(ny-1,i) = 4.
d(2,i)       = 0.5*b0*(f(i,3)-f(i,1))/ds
d(2,i)       = d(2,i) + 0.25*c0*(f(i,4)-f(i,2))/ds
d(ny-1,i) = 3.*(f(i,ny) - f(i,ny-2))/ds
!
do j=3,ny-2
  b(j,i) = a0
d(j,i) = 0.5*b0*(f(i,j+1)-f(i,j-1))/ds
d(j,i) = d(j,i) + 0.25*c0*(f(i,j+2)-f(i,j-2))/ds
end do
end do
!
```

```
do j=1,ny
cys(j)=c(j)
end do
cys(1)=0.0
!
call thomas(ny,nx,a,cys,b,d,cc)
!
do j=1,ny
do i=1,nx
f(i,j)=d(j,i)/hy(j)
  end do
end do
!
return
end
!------------------------------------------------------------
subroutine dpadyv(f)
!
!     Lele's Pade scheme 3/4/6 (6th order in interior)
!     for "image" variables [phi(+y)=-phi(-y)] of
symmetrical domain
!     (6th order at y=0)
!
include 'incl.for'
!
dimension f(nx,ny)
dimension a(ny),b(ny,nx),c(ny),d(ny,nx),cc(ny,nx),cyv(ny)
dimension y(ny),hy(ny),h2y(ny)
common /pade/ a0,b0,c0,a1,b1,c1
common /pady/ ds,a,c
common /mapy/ y,hy,h2y
!
do i=1,nx
!
b(1,i)    = a0
b(ny,i) = 2.
d(1,i)    = 0.5*b0*(f(i,2)+f(i,2))/ds
d(1,i)    = d(1,i) + 0.25*c0*(f(i,3)+f(i,3))/ds
d(ny,i) = (5.*f(i,ny) - 4.*f(i,ny-1) - f(i,ny-2))/ds
!
b(2,i)        = a0
```

```fortran
b(ny-1,i) = 4.
d(2,i)      = 0.5*b0*(f(i,3)-f(i,1))/ds
d(2,i)      = d(2,i) + 0.25*c0*(f(i,4)+f(i,2))/ds
d(ny-1,i) = 3.*(f(i,ny) - f(i,ny-2))/ds
!
do j=3,ny-2
b(j,i) = a0
d(j,i) = 0.5*b0*(f(i,j+1)-f(i,j-1))/ds
d(j,i) = d(j,i) + 0.25*c0*(f(i,j+2)-f(i,j-2))/ds
end do
end do
!
do j=1,ny
cyv(j)=c(j)
end do
cyv(1)=2.0
!
call thomas(ny,nx,a,cyv,b,d,cc)
!
do j=1,ny
do i=1,nx
  f(i,j)=d(j,i)/hy(j)
end do
end do
!
return
end
!-----------------------------------------------------------
subroutine d2padx(f)
!
!     Lele's Pade scheme for 2nd derivative 3/4/6 (6th
order in interior)
!
      include 'incl.for'
!
dimension f(nx,ny),df(nx,ny)
dimension a(nx),b(nx,ny),c(nx),d(nx,ny),cc(nx,ny)
dimension x(nx),hx(nx),h2x(nx)
common /pade2/ a0,b0,c0,a1,b1,c1
common /pad2x/ ds,a,c
common /mapx/ x,hx,h2x
!
```

```
do j=1,ny
do i=1,nx
df(i,j)=f(i,j)
end do
end do
!
call dpadex(df)
!
dsq=ds**2
!
do j=1,ny
!
b(1,j) =1.
b(nx,j)=1.
d(1,j) =(13.*f(1,j) - 27.*f(2,j) + 15.*f(3,j) - f(4,j))/
dsq
d(nx,j)=(13.*f(nx,j)-27.*f(nx-1,j)+15.*f(nx-2,j)-f(nx-
3,j))/dsq
!
b(2,j)   = 10.
b(nx-1,j) = 10.
d(2,j)   = 12.*(f(3,j) - 2.*f(2,j) + f(1,j))/dsq
d(nx-1,j) = 12.*(f(nx,j) - 2.*f(nx-1,j) + f(nx-2,j))/dsq
!
do i=3,nx-2
b(i,j) = a0
d(i,j) = b0*(f(i+1,j)-2.*f(i,j)+f(i-1,j))/dsq
d(i,j) = d(i,j) + 0.25*c0*(f(i+2,j)-2.*f(i,j)+f(i-2,j))/
dsq
end do
end do
!
call thomas(nx,ny,a,c,b,d,cc)
!
do j=1,ny
do i=1,nx
f(i,j)=d(i,j)/hx(i)**2-h2x(i)*df(i,j)/hx(i)**2
end do
end do
!
return
end
```

```
!-----------------------------------------------------------
subroutine d2pady(f)
!
!     Lele's Pade scheme for 2nd derivative 3/4/6 (6th
order in interior)
!
I       include 'incl.for'
!
dimension f(nx,ny),df(nx,ny)
dimension a(ny),b(ny,nx),c(ny),d(ny,nx),cc(ny,nx)
dimension y(ny),hy(ny),h2y(ny)
common /pade2/ a0,b0,c0,a1,b1,c1
common /pad2y/ ds,a,c
common /mapy/ y,hy,h2y
!
do j=1,ny
do i=1,nx
df(i,j)=f(i,j)
end do
end do
!
call dpadey(df)
!
dsq=ds**2
!
do i=1,nx
!
b(1,i) =1.
b(ny,i)=1.
d(1,i) =(13.*f(i,1) - 27.*f(i,2) + 15.*f(i,3) - f(i,4))/
dsq
d(ny,i)=(13.*f(i,ny)-27.*f(i,ny-1)+15.*f(i,ny-2)-f
(i,ny-3))/dsq
!
b(2,i)    = 10.
b(ny-1,i) = 10.
d(2,i)    = 12.*(f(i,3) - 2.*f(i,2) + f(i,1))/dsq
d(ny-1,i) = 12.*(f(i,ny) - 2.*f(i,ny-1) + f(i,ny-2))/dsq
!
do j=3,ny-2
b(j,i) = a0
d(j,i) = b0*(f(i,j+1)-2.*f(i,j)+f(i,j-1))/dsq
```

```fortran
d(j,i) = d(j,i) + 0.25*c0*(f(i,j+2)-2.*f(i,j)+f(i,j-2))/
dsq
end do
end do
!
call thomas(ny,nx,a,c,b,d,cc)
!
do j=1,ny
do i=1,nx
f(i,j)=d(j,i)/hy(j)**2-h2y(j)*df(i,j)/hy(j)**2
end do
end do
!
return
end
!-----------------------------------------------------------
subroutine d2pdys(f)
!
!      Lele's Pade scheme for 2nd derivative 3/4/6 (6th
order in interior)
!      for "twin" variables [phi(+y)=phi(-y)] of
symmetrical domain
!      (6th order at y=0)
!
include 'incl.for'
!
dimension f(nx,ny),df(nx,ny)
dimension a(ny),b(ny,nx),c(ny),d(ny,nx),cc(ny,nx),
c2ys(ny)
dimension y(ny),hy(ny),h2y(ny)
common /pade2/ a0,b0,c0,a1,b1,c1
common /pad2y/ ds,a,c
common /mapy/ y,hy,h2y
!
do j=1,ny
do i=1,nx
df(i,j)=f(i,j)
end do
end do
!
call dpadys(df)
!
```

```fortran
dsq=ds**2
!
do i=1,nx
!
b(1,i) =a0
b(ny,i)=1.
d(1,i) =b0*(f(i,2)-2.*f(i,1)+f(i,2))/dsq
d(1,i) =d(1,i)+ 0.25*c0*(f(i,3)-2.*f(i,1)+f(i,3))/dsq
d(ny,i)=(13.*f(i,ny)-27.*f(i,ny-1)+15.*f(i,ny-2)-f
(i,ny-3))/dsq
!
b(2,i)      = a0
b(ny-1,i) = 10.
d(2,i)      = b0*(f(i,3)-2.*f(i,2)+f(i,1))/dsq
d(2,i)      = d(2,i) + 0.25*c0*(f(i,4)-2.*f(i,2)+f(i,2))/dsq
d(ny-1,i) = 12.*(f(i,ny) - 2.*f(i,ny-1) + f(i,ny-2))/dsq
!
do j=3,ny-2
b(j,i) = a0
d(j,i) = b0*(f(i,j+1)-2.*f(i,j)+f(i,j-1))/dsq
d(j,i) = d(j,i) + 0.25*c0*(f(i,j+2)-2.*f(i,j)+f(i,j-2))/
dsq
end do
end do
!
do j=1,ny
c2ys(j)=c(j)
end do
c2ys(1)=2.0
!
call thomas(ny,nx,a,c2ys,b,d,cc)
!
do j=1,ny
do i=1,nx
f(i,j)=d(j,i)/hy(j)**2-h2y(j)*df(i,j)/hy(j)**2
end do
end do
!
return
end
!-----------------------------------------------------------
subroutine d2pdyv(f)
```

```
!
!       Lele's Pade scheme for 2nd derivative 3/4/6 (6th
order in interior)
!       for "image" variables [phi(+y)=-phi(-y)] of
symmetrical domain
!       (6th order at y=0)
!
include 'incl.for'
!
dimension f(nx,ny),df(nx,ny)
dimension a(ny),b(ny,nx),c(ny),d(ny,nx),cc(ny,nx),c2yv(ny)
dimension y(ny),hy(ny),h2y(ny)
common /pade2/ a0,b0,c0,a1,b1,c1
common /pad2y/ ds,a,c
common /mapy/ y,hy,h2y
!
do j=1,ny
do i=1,nx
df(i,j)=f(i,j)
end do
end do
!
call dpadyv(df)
!
dsq=ds**2
!
do i=1,nx
!
b(1,i) =a0
b(ny,i)=1.
d(1,i) =b0*(f(i,2)-2.*f(i,1)-f(i,2))/dsq
d(1,i) =d(1,i)+ 0.25*c0*(f(i,3)-2.*f(i,1)-f(i,3))/dsq
d(ny,i)=(13.*f(i,ny)-27.*f(i,ny-1)+15.*f(i,ny-2)-f
(i,ny-3))/dsq
!
b(2,i)      = a0
 b(ny-1,i) = 10.
d(2,i)      = b0*(f(i,3)-2.*f(i,2)+f(i,1))/dsq
d(2,i)      = d(2,i) + 0.25*c0*(f(i,4)-2.*f(i,2)-f(i,2))/
dsq
d(ny-1,i)  = 12.*(f(i,ny) - 2.*f(i,ny-1) + f(i,ny-2))/
dsq
```

```fortran
!
do j=3,ny-2
b(j,i) = a0
d(j,i) = b0*(f(i,j+1)-2.*f(i,j)+f(i,j-1))/dsq
d(j,i) = d(j,i) + 0.25*c0*(f(i,j+2)-2.*f(i,j)+f(i,j-2))/
dsq
   end do
end do
!
do j=1,ny
c2yv(j)=c(j)
end do
c2yv(1)=0.0
!
call thomas(ny,nx,a,c2yv,b,d,cc)
!
do j=1,ny
do i=1,nx
f(i,j)=d(j,i)/hy(j)**2-h2y(j)*df(i,j)/hy(j)**2
   end do
end do
!
return
end
!---------------------------------------------------------
!      tridiagonal solver
!      solves m systems of tridiagonal equations of size n
!      each system has identical sub- and super-diagonals
but
!      their diagonals and rhs's are different
!
subroutine thomas(n,m,a,c,b,d,cc)
!
dimension a(n),c(n),b(n,m),d(n,m),cc(n,m)
!
do j=1,m
v4=1./b(1,j)
d(1,j)=d(1,j)*v4
cc(1,j)=c(1)*v4
end do
!
do i=2,n-1
```

```
do j=1,m
  v2=b(i,j)-cc(i-1,j)*a(i)
v3=d(i,j)-d(i-1,j)*a(i)
v1=1./v2
d(i,j)=v3*v1
cc(i,j)=c(i)*v1
end do
end do
!
do j=1,m
v5=b(n,j)-cc(n-1,j)*a(n)
v6=d(n,j)-d(n-1,j)*a(n)
v7=1./v5
d(n,j)=v6*v7
end do
!
do j=1,m
do ii=1,n-1
i=n-ii
d(i,j)=d(i,j)-cc(i,j)*d(i+1,j)
end do
end do
!
return
end
!-----------------------------------------------------------
!      initialisation of Pade schemes
!
subroutine padeini
!
include 'incl.for'
dimension x(nx),hx(nx),h2x(nx),y(ny),hy(ny),h2y(ny)
dimension ax(nx),ay(ny),cx(nx),cy(ny)
dimension a2x(nx),a2y(ny),c2x(nx),c2y(ny)
logical cyl_flg,sym_flg,buo_flg
common /mapx/ x,hx,h2x
common /mapy/ y,hy,h2y
common /pade/ a0,b0,c0,a1,b1,c1
common /padx/ dsx,ax,cx
common /pady/ dsy,ay,cy
common /pade2/ a20,b20,c20,a21,b21,c21
common /pad2x/ ds2x,a2x,c2x
```

```fortran
common /pad2y/ ds2y,a2y,c2y
common /input1/ xl,yl,bx,by
common /input8/ cyl_flg,sym_flg,buo_flg
!
! ---set constants
!
dsx = 1./real(nx-1)
dsy = 2./real(ny-1)
if (sym_flg) dsy = 1./real(ny-1)
ds2x=dsx
ds2y=dsy
!
a0 = 3.
b0 = (4.*a0+2.)/3.
c0 = (4.-a0)/3.
a1 = 16.*(2.*a0+1.)/(40.-a0)
b1 = (4.*a1+2.)/3.
c1 = (4.-a1)/3.
!
a20 = 5.5
b20 = 4.*(a20-1.)/3.
c20 = (10.-a20)/3.
a21 = a20
b21 = b20
c21 = c20
!
! ---set arrays for x differentiation
!
do i=2,nx-1
ax(i)=1.
cx(i)=1.
end do
do i=2,nx-1
a2x(i)=1.
 c2x(i)=1.
end do
!
cx(1)=4.
ax(nx)=4.
c2x(1)=11.
a2x(nx)=11.
!
```

```
do i=1,nx
x(i)=xl*real(i-1)/real(nx-1)
hx(i)=xl
h2x(i)=0.
end do
!
! ---set arrays for y differentiation
!
do j=2,ny-1
ay(j)=1.
cy(j)=1.
end do
do j=2,ny-1
a2y(j)=1.
c2y(j)=1.
end do
!
cy(1)=4.
ay(ny)=4.
c2y(1)=11.
a2y(ny)=11.
!
!  Jiang & Luo's grid stretching (TCFD, 2000)
!
rc=1.0
bs=( log ((1.0+(exp(by)-1.0)*rc*2.0/yl)/ &
(1.0+(exp(-by)-1.0)*rc*2.0/yl)))/(2.0*by)
do j=1,ny
s=-1.+2.*real(j-1)/real(ny-1)
if (sym_flg) then
s=real(j-1)/real(ny-1)
y(j)=rc*(1.0+sinh(by*(s-bs))/sinh(by*bs))
hy(j)=rc*by*cosh(by*(s-bs))/sinh(by*bs)
h2y(j)=rc*by*by*sinh(by*(s-bs))/sinh(by*bs)
else
y(j)=0.5*yl*sinh(by*s)/sinh(by)
hy(j)=0.5*yl*by*cosh(by*s)/sinh(by)
h2y(j)=0.5*yl*by*by*sinh(by*s)/sinh(by)
end if
end do
!
return
end
```

In the above subroutines, the grid stretching was the one used by Jiang and Luo (2000). In using the above subroutines, an external file "incl.for" to define the array size must be established, which may look like

```
! 'include' file for grid numbers
parameter(nx=1351,ny=360)
! end of include
```

II. SECOND-ORDER CENTRAL DIFFERENTIATION

The second-order central difference has been broadly used in CFD and its formulation can be easily found (e.g., Anderson 1995). In the following sample subroutines, a discretized one-dimensional function g on a grid with the number of points given by nn=10 is considered. Subroutine dxc2 gives the output of the first and second derivatives using a three-point stencil, while subroutine dxstencil gives the output of the first and second derivatives using a five-point stencil. In both the subroutines, array xn represents the coordinate to be differentiated.

```
! --- Second order central differencing
!
  Subroutine dxc2
!
  parameter (nn=10)
  dimension xn(nn), derivc2(nn), deriv2c2(nn)
  common /x/ xn
  common /c2/ derivc2, deriv2c2
!
  h=1./real(nn-1.)
  hsq=h*h
!
! --- Second order forward and backward at boundaries
!       for first derivative
!
  derivc2(1) = (g(xn(1)+h)-g(xn(1)))/h
  derivc2(nn) = (g(xn(nn))-g(xn(nn)-h))/h
!
! --- Second order forward and backward at boundaries
!       for second derivative
!
  deriv2c2(1) = (g(xn(1)+2*h)-2*g(xn(1)+h)+g(xn(1)))/hsq
```

```
   deriv2c2 (nn) = (g(xn(nn)-2*h)-2*g(xn(nn)-h)+g(xn(nn)))/
hsq
!
   do i=2,nn-1
   derivc2 (i) = (g(xn(i)+h)-g(xn(i)-h))/(h+h)
   deriv2c2 (i) = (g(xn(i)+h)-2*g(xn(i))+g(xn(i)-h))/hsq
   end do
!
   return
   end
!
! --- Five-point stencil
!
   Subroutine dxstencil
!
   parameter (nn=10)
   dimension xn(nn), derivst(nn), deriv2st(nn)
   common /x/ xn
   common /c5/ derivst, deriv2st
!
   h=1./real(nn-1.)
   hsq=h*h
!
! --- First derivative with forward and backward schemes
!       at the boundaries with central scheme at the next to
!       the boundaries points
!
   derivst (1) = (g(xn(1)+h)-g(xn(1)))/h
   derivst (nn) = (g(xn(nn))-g(xn(nn)-h))/h
!
   derivst (2) = (g(xn(2)+h)-g(xn(2)-h))/(h+h)
   derivst (nn-1) = (g(xn(nn-1)+h)-g(xn(nn-1)-h))/(h+h)
!
! --- Second derivative with forward and backward schemes
       at the boundaries with central scheme at the next to
!       the boundaries points
!
   deriv2st (1) = (g(xn(1)+2*h)-2*g(xn(1)+h)+g(xn(1)))/hsq
   deriv2st (nn) = (g(xn(nn)-2*h)-2*g(xn(nn)-h)+g(xn(nn)))/hsq
!
   deriv2st (2) = (g(xn(2)+h)-2*g(xn(2))+g(xn(2)-h))/hsq
```

```
  deriv2st(nn-1)=(g(xn(nn-1)+h)-2*g(xn(nn-1))+g(xn
(nn-1)-h))/hsq
!
  do i=3,nn-2
  derivst(i)=(-g(xn(i)+2*h)+8*g(xn(i)+h)-8*g(xn(i)-h)&
  +g(xn(i)-2*h))/(12*h)
  deriv2st(i)=(-g(xn(i)+2*h)+16*g(xn(i)+h)-30*g(xn(i))&
  +16*g(xn(i)-h)-g(xn(i)-2*h))/(12*hsq)
  end do
!
  return
  end
```

REFERENCES

Anderson, John D. 1995. *Computational fluid dynamics: The basics with applications*. New York: McGraw-Hill.

Conte, Samuel D. and de Boor, Carl W. 1972. *Elementary numerical analysis*. New York: McGraw-Hill.

Jiang, X. and Luo, K. H. 2000. Direct numerical simulation of the puffing phenomenon of an axisymmetric thermal plume. *Theoretical and Computational Fluid Dynamics* 14: 55–74.

Lele, S.K. 1992. Compact finite-difference schemes with spectral-like resolution. *Journal of Computational Physics* 103: 16–42.

Index